CRC Handbook of Mathematical Curves and Surfaces

Author

David H. von Seggern
Senior Geophysicist
Exploration Division
Phillips Petroleum Company
Bartlesville, Oklahoma

CRC Press, Inc.
Boca Raton, Florida

Library of Congress Cataloging-in-Publication Data

von Seggern, David Henry
CRC handbook of mathematical curves and surfaces / author, David Henry von Seggern.
p. cm.
Bibliography: p.
Includes index.
ISBN 0-8493-0155-6
1. Curves on surfaces--Handbooks, manuals, etc. I. Title
QA643, V66 1990
516.3′5--dc20 89-33507
CIP

This book represents information obtained from authentic and highly regarded sources. Reprinted material is quoted with permission, and sources are indicated. A wide variety of references are listed. Every reasonable effort has been made to give reliable data and information, but the author and the publisher cannot assume responsibility for the validity of all materials or for the consequences of their use.

Direct all inquiries to CRC Press, Inc., 2000 Corporate Blvd., N.W., Boca Raton, Florida, 33431.

International Standard Book Number 0-8493-0155-6

Library of Congress Number 89-33507

Printed in the United States of America 2 3 4 5 6 7 8 9 0

PREFACE

Mathematical functions are a fundamental and prevalent ingredient in the endeavors of scientists and engineers. The conclusions, predictions, and analyses of such professionals are most often concisely contained in the abstract relations which are commonly referred to as "curves" when illustrated. Many special curves can be found in mathematical tables, such as the *CRC Handbook of Mathematical Sciences,*[1] and in mathematical dictionaries (for example, James and James[2]). The National Bureau of Standard's handbook (Abramowitz and Stegun[3]) is the acknowledged English-language source for special functions in physics and engineering. The recent work entitled *A Catalog of Special Plane Curves*[4] is an excellent source for illustrations of interesting functions in two dimensions. Yet, in spite of the frequent and widespread use of mathematical functions, there has been, to date, a lack of a single volume in which a diversity of curves appear in graphic form. Unexpectedly, there is an absence of any work which illustrates the spectrum of simple functions found in most integral tables. Thus, there is not a single reference work which draws together the entire gamut of forms which the modern scientist or engineer uses within a career. The need for such a reference volume is long overdue, especially in light of the fact that "curves" have become the ready tool of many other disciplines due to the computational and storage powers of modern electronic computers. Lastly, many of the curves appearing in older reference works show the imprecision of hand drafting methods; and the reappearance of familiar curves in precise, computer-plotted form should serve a useful purpose in itself.

"Curves" are abstractions of the form and motion of the physical world. Scientists have analyzed this physical world for millenia in order to render these abstract expressions, in the most minute detail, from gross astronomical movements to inifinitesimal atomic phenomena. It is now possible for a remarkably detailed synthesis of natural phenomena to be created by the proper use of these abstractions. Some such synthetic renditions have emerged from the field of computer graphics already: mountainous terrain, cloud formations, trees, to name a few, nearly indistinguishable from reality. Modern scientists' skillful, mathematical description of the motion of nature, coupled with modern computing power, has also enabled them to make increasingly accurate predictions of natural events, such as for weather, earthquakes, and oceanic currents. All such endeavors involve, as the quantitative basis, functions whose "curves" are the visual representation of the predicted motion. Scientists and engineers can use this reference work in two ways to aid their work. In the "forward" manner, they can look up the equation of interest and see the corresponding visual form of the curve. In the "inverse" manner, they can select a particular curve visually to serve in data fitting or in computer modeling exercises.

This handbook, however, purports to serve a larger audience than those engaged in mathematics, science, and engineering. Architects, designers, draftsmen, and artists should benefit from this reference book of curves. New expressions of form can be imagined through even a casual scanning of the contents of this volume. And, if a general notion of the desired appearance exists, the appropriate curve can be located in this volume and its mathematical expression noted. The mathematical expressions given here can be readily translated into high-level programing languages (for example, FORTRAN) in order to generate a given curve in a particular environment of application. Recent graphics languages enable "cells", "segments", or "symbols" to be created once and stored for future use. These abstractions, which can be composed of one or more curve segments, may be placed in a computer-based design at any scale or rotation angle to achieve the desired effect. The computer revolution indeed makes curve generation easy and rapid and eliminates the former laborious hand calculations necessary to graph even the simplest curves. Achieving the most intricate and subtlest abstract forms, as well as the simple and plain, is possible for those who have only a rudimentary programing knowledge. The properly designed computer programs could open up this possibility even to those who have no grasp of the underlying equations.

This work is intended to contain all curves in common use in applied mathematics. In order to be comprehensive, the notion of "curve" has been extended beyond its usual connection with algebraic or transcendental functions. Here "curve" implies any line or surface in two or three dimensions which can be generated by a rule or set of rules expressible in mathematical terms. Such rules may be entirely smooth and deterministic; and the first part of this handbook is devoted to the curves represented in this way: algebraic forms, transcendental forms, and special integrals. Here the mathematicians, scientists, and engineers will find those curves familiar to them. Selection of functions for curve-fitting can be eased by use of this handbook, and questions concerning the form of a given function can be quickly settled. Those with careers in design can find curves here appropriate to their design goals. The latter part of this handbook comprises curves and surfaces which are not smoothly generated by a single relation, such as piecewise continuous functions, polygons, and polyhedra. When the generating rules include random components, a new series of curves and surfaces emerges—the subject of a future volume. The need for cataloging such curves is attributed to the influence on science of Mandelbrot[5] who has shown that the study and description of the random component of natural phenomena is as important, if not more important, than that of the deterministic component. Also, a future volume has been reserved to collect together many interesting and unusual curves which are not normally considered in pure mathematics. These curves will be most useful to artists and designers who are able to employ modern computer-assisted art and drafting systems.

This handbook begins with a chapter containing a qualitative summary of deterministic curve properties and a classification summary of such curves. An explanation of the means and conventions of presentation in later chapters is also given here. The first chapter is meant to acquaint the reader with fundamental mathematical properties of curves in order that application of the material of this handbook can be more knowledgeable and meaningful. Those with a solid background of calculus will find little new information here. A section on matrix transformations has been included to indicate how a given curve can be made to appear in many different visual forms. The following chapters are organized such that similar curves are grouped together for easy reference. Early chapters deal with curves in two dimensions, progressing from the simple to the complex. Later chapters extend the notion of curve to three dimensions, in the form of surfaces. Final chapters deal with piecewise continuous functions in two and three dimensions.

REFERENCES

1. **Beyer, W. H., Ed.,** *CRC Standard Mathematical Tables,* CRC Press, Boca Raton, 1978.
2. **James, G. and R. C. James, Eds.,** *Mathematics Dictionary,* Van Nostrand, New York, 1949.
3. **Abramowitz, M. and I. A. Stegun, Eds.,** Handbook of Mathematical Functions, National Bureau of Standards, Department of Commerce, Washington, D. C., 1964.
4. **Lawrence, J. D.,** *A Catalog of Special Plane Curves,* Dover Publications, New York, 1972.
5. **Mandelbrot, B. B.,** *The Fractal Geometry of Nature,* W. H. Freeman, San Francisco, 1983.

THE AUTHOR

David H. von Seggern, Ph.D., is a geophysicist currently with Phillips Petroleum Company of Bartlesville, Oklahoma. He previously worked for Teledyne Geotech in Alexandria, Virginia, on numerous aspects of underground nuclear test detection. During that time, he authored or co-authored 11 professional papers and nearly 40 company reports on the subject. He completed his education at the Pennsylvania State University with a dissertation on earthquake prediction which included an early application of fractal theory in seismology.

At Phillips Petroleum Company, Dr. von Seggern has specialized in applying computer graphics to the problems of processing and interpreting seismic data, has promoted seismic modeling as an aid in data interpretation, and has done research in seismic imaging methods using supercomputer technology.

TABLE OF CONTENTS

Chapter 2

Algebraic Curves

Chapter 1

INTRODUCTION

1.1. CONCEPT OF A CURVE

Let E^n be the Euclidean space of dimension n. (According to this definition, E^1 is a line, E^2 is a plane, and E^3 is a volume.) A ''curve'' in n-space is defined as the set of points which result when a mapping from E^1 to E^n is performed. In this reference work, only curves in E^2 and E^3 will be considered. Let t represent the independent variable in E^1. An E^2 curve is then given by

$$x = f(t); \quad y = g(t)$$

and an E^3 curve by

$$x = f(t); \quad y = g(t); \quad z = h(t)$$

where f, g, and h mean ''function of''. The domain of t is usually $(0,2\pi)$, $(-\infty,\infty)$, or $(0,\infty)$. These are the ''parametric'' representations of a curve. However, in 2-space, curves are commonly expressed as

$$y = f(x)$$

or as

$$f(x,y) = 0$$

which are the explicit and implicit forms, respectively. The explicit form is reducible from the parametric form when $x = f(t) = t$ in 2-space and when $x = f(t) = t$ and $y = g(t) = t$ in 3-space. The implicit form of a curve will often comprise more points than a corresponding explicit form. For example $y^2 - x = 0$ has two ranges in y, one positive and one negative, while the explicit form derived from solving the above equation gives $y = x^{1/2}$ for which the range of y is positive only.

Generally, the definition of a curve imposes a smoothness criterion,[1] meaning that the trace of the curve has no abrupt changes of direction (continuous first derivative). However, for purposes of this reference work, a broader definition of curve is proposed. Here, a curve may be composed of smooth branches, each satisfying the above definition, provided that the intervals over which the curve branches are distinctly defined are contiguous. This definition will encompass forms such as polygons or sawtooth functions.

1.2. CONCEPT OF A SURFACE

This reference work will treat only surfaces in 3-space (E^3). Therefore a "surface" is defined as the mapping from E^2 to E^3 according to

$$x = f(s,t); \quad y = g(s,t); \quad z = h(s,t)$$

As for curves, the conversion from this parametric form to more common forms

$$z = f(x,y)$$

or

$$f(x,y,z) = 0$$

may not be possible in some cases. Again, a smoothness criterion[1] is desirable; but the generalized definition of surface requires that this smoothness criterion only be satisfied piecewise for all distinct mappings of the (s,t) plane over which the surface is defined. These generalized surfaces are termed "manifolds", and examples such as cubes fall into the class of surfaces which can be defined in a deterministic manner.

1.3. COORDINATE SYSTEMS

The number of available coordinate systems for representing curves is large and even larger for surfaces. However, to maintain uniformity of presentation throughout this volume, only the following will be used:

2-D	**3-D**
Cartesian, polar	Cartesian, cylindrical, spherical

The term "parametric" is often used as though it were a coordinate system, but it is really a "representation" of coordinates in terms of an additional independent parameter which is not itself a coordinate of the E^n space in which the curve or surface exists.

1.3.1. Cartesian Coordinates

The Cartesian system is illustrated in Figure 1 for two dimensions. This is the most natural, but not always the most convenient, system of coordinates for curves in two dimensions. Coordinates of a point p are measured linearly along two axes which intersect with a right angle at the origin (0,0). The Cartesian system is also called the "rectangular" system. For three dimensions, an additional axis, orthogonal to the other two, is placed as shown in Figure 2.

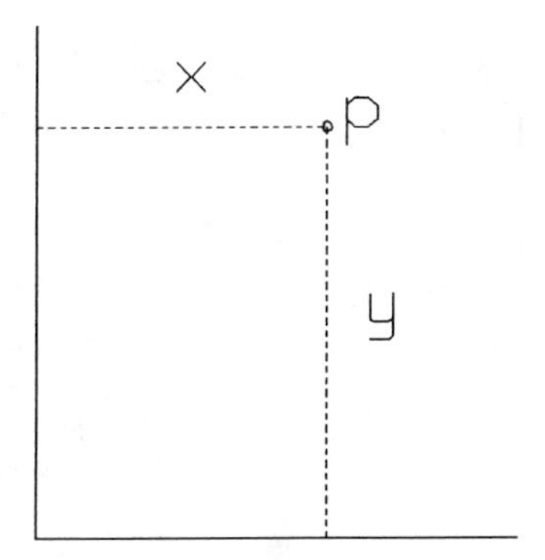

FIGURE 1. The Cartesian coordinate system for two dimensions.

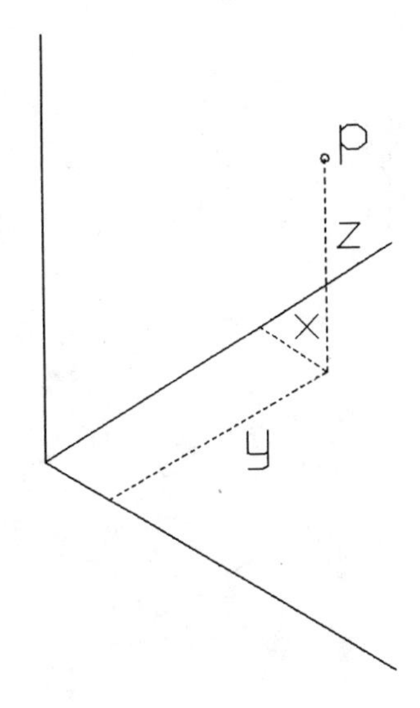

FIGURE 2. The Cartesian coordinate system for three dimensions.

1.3.2. Polar Coordinates

Polar coordinates (r,θ) are defined for two dimensions and are a desirable alternative to Cartesian ones when the curve is point symmetric and exists only over a limited domain and range of the variables x and y. As illustrated in Figure 3, the coordinate r is the distance of the point p from the origin and the coordinate θ is the counterclockwise angle which the line from the origin to p makes with the horizontal line through the origin to the right. Clockwise rotations are measured in negative θ relative to this line. Transformations from polar to Cartesian, and vice versa, are made according to:

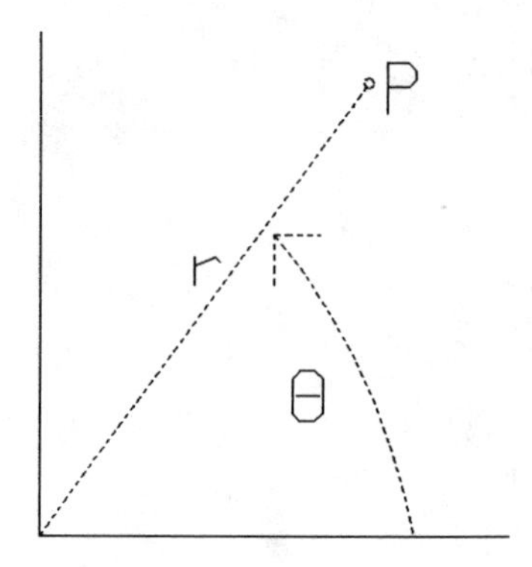

FIGURE 3. The polar coordinate system for two dimensions.

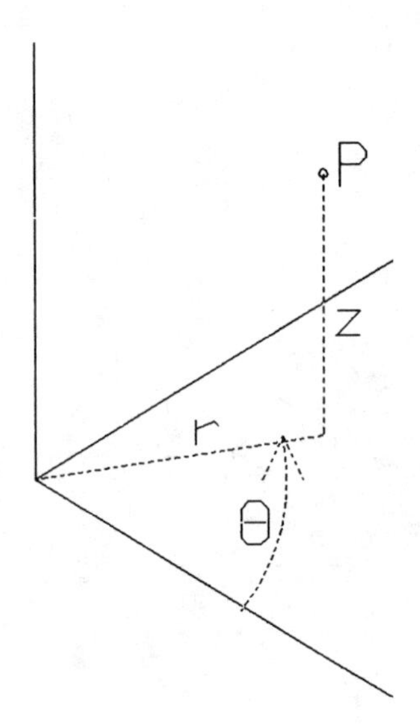

FIGURE 4. The cylindrical coordinate system for three dimensions.

$$x = r \cdot \cos(\theta); \quad y = r \cdot \sin(\theta)$$

$$r = (x^2 + y^2)^{1/2}; \quad \theta = \arctan(y/x)$$

1.3.3. Cylindrical Coordinates

Cylindrical coordinates are used in three dimensions. They combine the (r,θ) polar coordinates of two dimensions with the third coordinate z measured perpendicularly from the x-y plane at (r,θ) to the point p at (r,θ,z) as in Figure 4. The normal convention is for z to be positive upward. Transformation from cylindrical to Cartesian coordinates involves only the polar-to-Cartesian transformations given above because the z coordinate is unchanged. Cylindrical coordinates are often

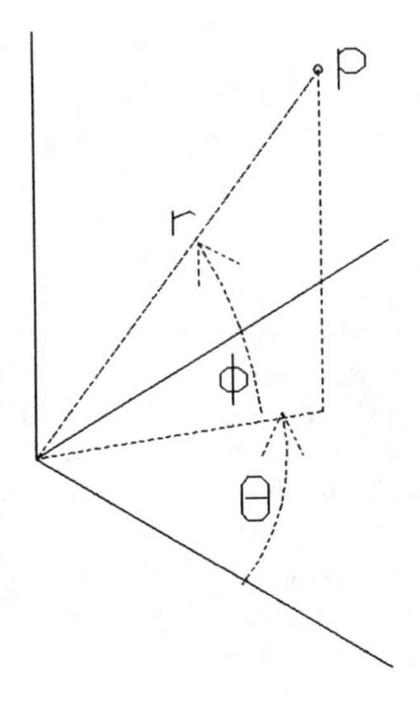

FIGURE 5. The spherical coordinate system for three dimensions.

appropriate when surfaces are axially symmetric about the z axis; for example, in representing the form $r^2 = z$.

1.3.4. Spherical Coordinates

As illustrated in Figure 5, let a point in E^3 lie at a radial distance r along a vector from the origin. Project this vector to the x-y plane and let the angle between the vector and its projection be ϕ. Now measure the angle θ of the projected line in the x-y plane as for polar coordinates. Then (r,θ,ϕ) are the spherical coordinates of p. The transformations from spherical to Cartesian coordinates, and vice versa, are given by:

$$x = r \cdot \cos(\theta) \cdot \sin(\phi); \quad y = r \cdot \sin(\theta) \cdot \sin(\phi); \quad z = r \cdot \cos(\phi)$$

$$r = (x^2 + y^2 + z^2)^{1/2}; \quad \theta = \arctan(y/x); \quad \phi = \arctan[(x^2 + y^2)^{1/2}/z]$$

Spherical coordinates are often appropriate for surfaces having point symmetry about the origin. The usual coordinates of geography, which refer to points on the earth by latitude and longitude, are a spherical system.

1.4. QUALITATIVE PROPERTIES OF CURVES AND SURFACES

Curves and surfaces exhibit a wide variety of forms. Particular attributes of form are derivable from the equations themselves, and many texts treat these in rigorous detail. The purpose here is not to duplicate such explicit and analytical treatment but rather to present the properties of curves and surfaces in a ''qualitative'' manner to which their visible forms are naturally and easily related. Understanding of these properties enables one to choose the appropriate curve for a given purpose (for example, data fitting) or to modify, when necessary, an equation given in this volume into one more suitable for a given purpose.

1.4.1. Derivative

A fundamental quantity associated with a curve, or function, is the derivative, which exists at all continuous points of the curve (except singular points as described in Section 1.4.7). Although the definition of derivative can be made with analytical rigor[1], in graphical terms the *derivative* at any point is the slope of the tangent line at that point and is written as dy/dx for two-dimensional curves. For three-dimensional curves, the tangent line is along the trajectory of the curve, and three such derivatives are possible using the three pairs of x, y, and z coordinates. Closely associated with the derivative is a curve's *normal* which is the line perpendicular to the tangent. In two dimensions the normal is a single line, but in three dimensions the normal swings in a full circle about the tangent to the curve.

As for curves, the derivative of a surface is a fundamental quantity. The derivative at any continuous point of a surface relates to the tangent plane of the surface at that point. For this plane, three "partial" derivatives exist, written as $\delta y/\delta z$, $\delta z/\delta x$, and $\delta x/\delta y$ (or their inverses), which are the slopes of the lines formed at the intersection of the tangent plane with the y-z, z-x, and x-y planes, respectively. The normal to the surface at a point is the vector orthogonal to the surface there. It is defined at all points for which the surface is smooth by the partial derivatives

$$n_p = \left[\begin{pmatrix}\delta y/\delta s & \delta y/\delta t\\ \delta z/\delta s & \delta z/\delta t\end{pmatrix}, \begin{pmatrix}\delta z/\delta s & \delta z/\delta t\\ \delta x/\delta s & \delta x/\delta t\end{pmatrix}, \begin{pmatrix}\delta x/\delta s & \delta x/\delta t\\ \delta y/\delta s & \delta y/\delta t\end{pmatrix}\right]_p$$

using the parametric representation equations. If the surface can be expressed in the implicit form $f(x,y,z) = 0$, then simply

$$n_p = \left[\frac{\delta f}{\delta x}, \frac{\delta f}{\delta y}, \frac{\delta f}{\delta z}\right]_p$$

The above definitions give the (x,y,z) components of the normal vector, and it is customary to normalize them to (x',y',z') by dividing them with $(x^2 + y^2 + z^2)^{1/2}$ so that $x'^2 + y'^2 + z'^2 = 1$.

1.4.2. Symmetry

For curves in two dimensions, if

$$y = f(x) = f(-x)$$

holds, then the curve is *symmetric* about the y axis. The curve is *antisymmetric* about the y axis when

$$y = f(x) = -f(-x)$$

A simple example is powers of x: $y = x^n$. If n is even, the curve is symmetric; if n is odd, it is antisymmetric. Antisymmetry is also referred to as "symmetric with respect to the origin" or "point symmetric" about $(x,y) = (0,0)$.

For surfaces, three kinds of symmetry exist: point, axial, and plane. A surface has *point symmetry* when

$$z = f(x,y) = -f(-x,-y)$$

Simple examples of point symmetry are spheres or ellipsoids. A surface has *axial symmetry* when

$$z = f(x,y) = f(-x,-y)$$

An example of axial symmetry is a paraboloid. Finally, a surface has *plane symmetry* about the (y,z) plane when

$$z = f(x,y) = f(-x,y)$$

Similarly, symmetry about the (x,z) plane implies

$$z = f(x,y) = f(x,-y)$$

Examples of plane symmetry include $z = xy^2$ and $z = e^x\cos(y)$.

1.4.3. Extent

The extent of a curve is defined by the range (y variation) and domain (x variation) of the curve. The extent is *unbounded* if both x and y values can extend to infinity (for example, $y = x^2$). The extent is *semibounded* if either y or x has a bound less than infinity. The transcendental equation $y = \sin(x)$ is such a curve because the range is limited between negative and positive unity. A curve is *fully bounded* if both x and y bounds are less than infinity. A circle is a simple example of this type of extent.

For surfaces the concept of extent can be applied in three dimensions where domain applies to x and y while range applies to z. Surfaces formed by revolution of a curve in the (y,z) or (x,z) plane about the z axis will possess the same extent property that the two-dimensional curve had. For example, an ellipse in the (x,z) plane gives an ellipsoid as the surface of revolution—both have the fully bounded property. Similarly, any surface formed by continuous translation of a two-dimensional curve (for example, a parabolic sheet) will have the same extent property as the original curve.

1.4.4. Asymptotes

The *y asymptotes* of a curve are defined by

$$y_a = \lim_{x \to \pm\infty} f(x)$$

Although this definition includes asymptotes at infinity, only those with $|y_a|<\infty$ are of interest. Asymptotic values are often crucial in choosing and applying functions. Physically, an equation may or may not properly describe real phenomena, depending on its asymptotic behavior. Note that even though a curve may be semibounded, its asymptote may not be determinable. An example of a semibounded function with a y asymptote is $y = e^{-x}$ while one without an asymptote is $y = \sin(x)$.

The *x asymptotes* of a curve may be defined in a similar manner:

$$x_a = \lim_{y \to \pm\infty} f(y)$$

when the function is inverted to give $x = f(y)$. An example of a curve with a finite x asymptote is $y = (c^2 - x^2)^{1/2}$ whose asymptote lies at $x = +c$ or $x = -c$.

In addition, curves may have asymptotes that are any arbitrary lines in the plane, not simply horizontal or vertical lines; and the limiting requirements are similar to the forms given above for horizontal or vertical asymptotes. For instance, the equation $y = x + 1/x$ has $y = x$ as its asymptote.

1.4.5. Periodicity

A curve is defined as *periodic* on x with period X if

$$y = f(x + nX)$$

is constant for all integers n. The transcendental function $y = \sin(ax)$ is an example of a periodic curve. A polar coordinate curve can also be defined as periodic with period α in terms of angle θ if

$$r = f(\theta + n\alpha)$$

is constant for all integers n. An example of such a periodic curve is $r = \cos(4\theta)$ which exhibits eight "petals" evenly spaced around the origin.

Surfaces are periodic on x and y with periods X and Y, respectively, if

$$z = f(x + nX, y + mY)$$

is constant for all integers n and m. A surface also may be periodic in only x or only y. A cylindrical-coordinate surface may be periodic with period α in terms of the angle θ if

$$z = f(r, \theta + n\alpha)$$

is constant for all integers n. Another type of periodicity expressible in cylindrical coordinates is in the radial direction with period R, when

$$z = f(r + nR,\theta)$$

is constant for all integers n. An example of such periodicity is given by $z = \cos(2\pi r)\cos(\theta)$, which has a period of unity in r.

1.4.6. Continuity

A curve is *continuous* at a point x_0, provided it is defined at x_0, when

$$y^+ = \lim_{x \to x_0^+} f(x)$$

and

$$y^- = \lim_{x \to x_0^-} f(x)$$

are finite and equal. Here "+" and "−" refer to approaching x_0 from the right and left, respectively. Discontinuities may be finite or infinite: the former implies $y^+ \neq y^-$ even though they are both finite while the latter implies one or both limits are infinite. For surfaces, the paths to a point $p_0 = (x_0,y_0)$ are infinite in number; and continuity exists only if the surface is defined at p_0 and

$$z = \lim_{p \to p_0} f(p)$$

is constant for all possible paths. When the curve or surface is undefined at x_0 or p_0 and the above relations hold, it is said to be discontinuous, but with a "removable discontinuity". For any points at which the above relations do not hold, the curve or surface is discontinuous, with an "essential discontinuity" at such points. The curve $y = \sin(x)/x$ has a removable discontinuity and is therefore continuous in appearance while $y = 1/x$ has an essential discontinuity at $x = 0$ and is therefore discontinuous in appearance. Curves and surfaces are "differentiable" (meaning the derivative exists) everywhere that they are either continuous or have removable discontinuities.

1.4.7. Singular Points

Curves and surfaces given by polynomials of degree 3 or greater may contain singular points. Writing the function for a two-dimensional curve as

$$f(x,y) = 0$$

the derivative dy/dx can be written as

$$\frac{dy}{dx} = \frac{g(x,y)}{h(x,y)}$$

where g and h are functions of x and y. If for a given point p(x,y) the functions g and h both vanish, the derivative becomes the indeterminate form 0/0; and p(x,y) is then a *singular point* of the curve. Singular points imply that two or more branches of the curve meet or cross. If two branches are involved, it is a double point; if three are involved, it is a triple point; and so on. Singularities at triple or higher points are not as commonly encountered as those at double points. Double-point singularities for two-dimensional curves are classified as follows:

1. "Isolated" (or "conjugate") points are where a single point is disjoint from the remainder of the curve. In this case, the derivative is imaginary.
2. "Node" points are where the two derivatives are real and unequal, such that the curve crosses itself.
3. "Cusp" points are where the derivatives of two arcs become equal and the curve ends at this point. A "cusp of the first kind" involves second derivatives of opposite sign, and a "cusp of the second kind" involves second derivatives of the same sign.
4. "Double cusp" (or "osculation") points are where the derivatives of two arcs become equal while the two arcs of the curve are continuous along both directions away from such points. Double cusps may also be of the first or second kind, as for single cusps.

Curves having one or more nodes will exhibit "loops" which enclose areas. Curves having osculations may also exhibit loops, on one or both sides of the osculation point.

The concept of singular points is extendable to surfaces. Many surfaces are the result of the revolution of a two-dimensional curve about some line; such surfaces retain the singular points of the curve, except that each such point on the curve, unless on the axis of revolution, becomes a circular ring of singular points centered on the axis of revolution. Singular points appear on more complicated surfaces also, but an analysis of the possibilities is beyond the scope of this volume.

1.4.8. Critical Points

Points of a curve $y = f(x)$ at which the derivative $dy/dx = 0$ are termed *critical points*, of which there are three types:

1. "Maximum" points are where the curve is concave downward and thus the second derivative $d^2y/dx^2 > 0$.
2. "Minimum" points are where the curve is concave upward and thus the second derivative $d^2y/dx^2 < 0$.
3. "Inflection" points are where $d^2y/dx^2 = 0$ and the curve changes its direction of concavity.

For surfaces $z = f(x,y)$, the critical points lie at $\delta z/\delta x = \delta z/\delta y = 0$. Maximum and minimum points of surfaces are defined similar to those of curves, except both second derivatives must together be greater than or less than zero. In the case that they are of opposite sign, the critical point is termed a "saddle." Such critical points are "nondegenerate"[2] and are isolated from other critical points. More complicated types of critical points occur for surfaces and are classified as "degenerate" or "nondegenerate", depending on whether the determinant of

$$\begin{pmatrix} \delta^2 z/\delta x^2 & \delta^2 z/\delta x \delta y \\ \delta^2 z/\delta x \delta y & \delta^2 z/\delta y^2 \end{pmatrix}$$

vanishes or not, respectively. The surface $z = x^2 + y^2$ has a single nondegenerate critical point while $z = x^2y^2$ has two continuous lines of degenerate critical points, intersecting at (0,0).

1.4.9. Zeroes

The *zeroes* of a two-dimensional function f(x) occur where $y = f(x) = 0$ and are isolated points on the x-axis. (For polynomial functions, the zeroes are often referred to as the "roots".) Similarly, the zeroes of a three-dimensional function f(x,y) occur where $z = f(x,y) = 0$; but the locus of these points form one or more distinct, continuous curves in the x-y plane. The zeroes of certain functions are important in characterizing their oscillatory behavior (for example, the function sin(x)) while the zeroes of other functions may be unique points of interest in physical applications. Not all functions, as defined, have zeroes; for example, the function $f(x) = 2 - \cos(x)$ has unity as its lower bound. However, such a function can be translated in one or the other y direction to produce a function having zeroes in addition to all the qualitative properties of the original function.

The definition of the exact zeroes of a function is often difficult and often must be accomplished by numerical methods on a computer. Zeroes of many functions are tabulated in standard references such as Abramowitz and Stegun.[3]

1.4.10. Integrability

The function $y = f(x)$ defined over the interval [a,b] has the integral

$$I = \int_a^b y \, dx$$

The integral exists if I converges to a single, bounded value for a given interval; and the function is said to be *integrable*. Note that I may exist under two abnormal circumstances:

1. Either a or b, or both, extend to infinity.
2. The function y has an infinite discontinuity at one or both endpoints or at one or more points interior to [a,b].

Under either of these circumstances, the integral is an "improper integral". Proving the existence of the integral of a given function is not always straightforward, and a discussion is beyond the scope of this volume.

"Transient" functions always have an integral on the interval $[0,\infty]$ and are often given as solutions to physical problems in which the response of a medium to a given input or disturbance is sought. Such responses must possess an integral if the input was finite and measurable. Examples of such functions are $y = e^{-ax}\sin(bx)$ or $y = 1/(1 + x^2)$.

Surfaces are integrable when

$$I = \int_a^b \int_c^d z \, dx \, dy$$

exists. Improper integrals of surfaces are defined in the same manner as those of two-dimensional curves. Transient responses exist for three dimensions and are integrable also.

A curve property which has an important consequence for integration is that of "even" and "odd" functions. Even functions have $f(x) = f(-x)$, and for such curves

$$I = 2\int_0^a f(x)dx$$

if I exists over $[-a,a]$. For odd functions $f(x) = f(-x)$, and $I = 0$ over any interval $[-a,a]$. This concept can be easily extended to surfaces.

1.4.11. Multiple Values

A curve is *multivalued* if, for a given x, y has two or more distinct values. A simple example is $y^2 = x$. Multivalued functions are not integrable in the normal sense although one or more particular branches of the curve may have well-defined integrals.

While a curve may be multivalued in its Cartesian-form equation, the polar form of the equation may be single-valued, in the sense that only one value of r exists for each value of angle θ. Compare, for example,

$$(x^2 + y^2)^3 = (x^2 - y^2)^2$$

which is the equation of a quadrifolium, with its polar equation

$$r = \cos(2\theta)$$

Integrability is affected by the choice of coordinate system; this example shows that, when an integral is not defined due to a function being multivalued, it may be well defined when the transformation to polar coordinates is made and the integral evaluated along the polar angle θ.

Similarly, surfaces may be single- or multivalued depending upon whether z takes on one or more values for a given (x,y) point.

1.4.12. Curvature

Given that a unit of length along the curve path is ds and that the tangent line changes its direction over ds by an angle $d\theta$ where θ is the angle of the tangent with the x-axis, then the *radius of curvature* is given by

$$\rho = \left|\frac{ds}{d\theta}\right|$$

This radius can be expressed in terms of the derivatives of the curve also. If the curve is expressed implicitly as $f(x,y) = 0$ and if f_x and f_y are the first partial derivatives and f_{xx}, f_{yy}, and f_{xy} are the second partial derivatives, then

$$\rho = \frac{(f_x^2 + f_y^2)^{3/2}}{f_{xx}f_y^2 - 2f_{xy}f_xf_y + f_{yy}f_x^2}$$

When the curve is expressed in polar coordinates and the derivatives $dr/d\theta$ and $d^2r/d\theta^2$ are given by r' and r'' respectively, then the radius of curvature is

$$\rho = \frac{(r^2 + r'^2)^{3/2}}{r^2 + 2r'^2 - rr''}$$

The radius of curvature at lobes of polar curves is of interest in order to define the "tightness" of the lobes. At the peak of the lobe, $r' = 0$ and $\rho = r^2/(r - r'')$. This reduces to $\rho = r$ in the case of a circle, for which $r'' = 0$.

Using the same formula as for curves above, curvature of surfaces can be measured along any arbitrary linear arc of the surface made by an intersecting plane, where θ would be the angle of the tangent line relative to the horizontal in the intersecting plane. Thus the curvature of a surface is relative to the perspective it is viewed from.

1.5. CLASSIFICATION OF CURVES AND SURFACES

The family of two- and three-dimensional curves can be illustrated in Figure 6. This particular schematic reflects the organization of this reference work, and every curve which can be traced by a given mathematical equation or given set of mathematical rules can be placed in one of the categories shown. There is a top-level dichotomy between determinate and random curves, but no further reference will be made to random curves in this volume. A "determinate" curve is one for which the functional relationship between x and y is known everywhere from the equation or set of rules in the abstract. No realization is required to produce the curve, for it is contained wholly within its defining equations or rules. On the other hand, a random curve will have a random factor or term in its mathematical definition such that an actual realization is required to produce the curve, which will differ from any other realization. For example, $y = \sin(x) + w(x)$ where $w(x)$ is a random variable on x, defines a random curve.

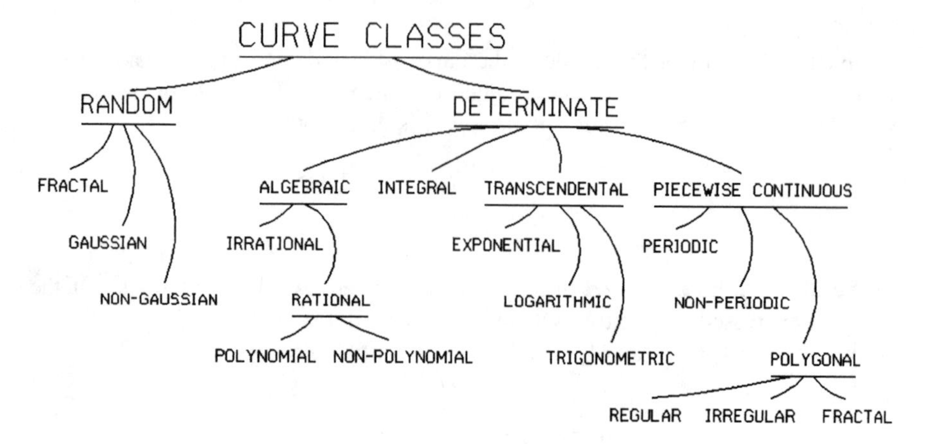

FIGURE 6. A classification of curves and surfaces for this handbook.

At the second level in Figure 6, the distinction is made between algebraic, transcendental, integral, and nondifferentiable curves as described below.

1.5.1. Algebraic Curves

A *polynomial* is defined as a summation of terms composed of integer powers of x and y. An *algebraic* curve is one whose implicit function

$$f(x,y) = 0$$

is a polynomial in x and y (after rationalization, if necessary). Because a curve is often defined in the explicit form

$$y = f(x)$$

there is a need to distinguish rational and irrational functions of x. A "rational" function of x is a quotient of two polynomials in x, both having only integer powers. An "irrational" function of x is a quotient of two polynomials, one or both of which has a term (or terms) with power p/q, where p and q are integers. Irrational functions can be rationalized, but the curves will not be identical before and after rationalization. In general, the rationalized form has more branches; for example, consider $y = x^{1/2}$ which is rationalized to $y^2 = x$. The former curve has only one branch (for positive y) if a strict definition of the radical is used whereas the latter has two branches, for $y < 0$ and $y > 0$. In this reference work, the rationalized curve will be presented graphically in all cases, even though the curve is printed in its irrational form for simplicity.

Besides simple polynomials, rational functions are often grouped into sets convenient for certain mathematical applications. Examples of such polynomial sets are: Chebychev polynomials, Laguerre polynomials, and Bernoulli polynomials. Most polynomial sets have the property of "orthogonality", meaning that for any two functions of the set

$$\int w(x) \cdot f_1(x) \cdot f_2(x) \cdot dx = 0$$

over the defined domain of x for the particular set, where w(x) is a weighting function. This property ensures that the different curves within the set make a distinct contribution to the set.

1.5.2. Transcendental Curves

A *transcendental* curve is one which cannot be expressed as an infinite polynomial in x and y. These are curves containing one or more of the following forms: exponential (e^x), logarithmic (log(x)), or trigonometric (sin(x), cos(x)). (The hyperbolic functions are often mentioned as part of this group, but they are not really distinct because they are forms composed of exponential functions.) Any curve expressed as a mixture of transcendentals and polynomials is considered to be transcendental. All of the primary transcendental functions can, in fact, be expressed as infinite polynomial series:

$$e^x = \sum_{n=0}^{\infty} \frac{x^n}{n!} \qquad (-\infty < x < \infty)$$

$$\cos(x) = \sum_{n=0}^{\infty} \frac{(-1)^n x^{2n}}{(2n)!} \qquad (-\infty < x < \infty)$$

$$\sin(x) = \sum_{n=0}^{\infty} \frac{(-1)^n x^{2n+1}}{(2n+1)!} \qquad (-\infty < x < \infty)$$

$$\log(x) = 2\sum_{n=1}^{\infty} \frac{1}{2n-1} \left(\frac{x-1}{x+1}\right)^{2n+1} \qquad (x > 0)$$

1.5.3. Integral Curves

Certain continuous curves not expressible in algebraic or transcendental forms are familiar mathematical tools. These curves are equal to the integral of algebraic or transcendental curves by definition; examples include Bessel functions, Airy integrals, Fresnel integrals, and the error function. The integral curve is given by

$$y(b) = \int_a^b f(x)dx$$

where the lower limit of integration *a* is usually a fixed point such as $-\infty$ or 0. Like transcendental curves, these integral curves also have expansions in terms of power series or polynomial series, often making evaluation rather straightforward on computers.

1.5.4. Piecewise Continuous Functions

Members of the previous classes of curves (algebraic, transcendental, and integral) all have the property that (except at a few points, called singular points) the curve is smooth and differentiable. In the spirit of a broad definition of curve, a class of nondifferentiable curves appears in Figure 6. These curves have discontinuity of the first derivative as a basic attribute and are quite often composed of straight-line segments. Such curves include the simple polygonal forms as well as the intricate "regular fractal" curves of Mandelbrot[4] and the "turtle" tracks described in Hayes.[5]

1.5.5. Classification of Surfaces

In general, surfaces may follow the same classification scheme as curves (Figure 6). Many commonly used surfaces are either rotations of two-dimensional curves about an axis, thus giving axial, or possibly point, symmetry. In this case the independent variable x of the two-dimensional curve's equation can be replaced with the radial variable $r = (x^2 + y^2)^{1/2}$ to form the equation of the surface. Other commonly used surfaces are merely a continuous translation of a given two-dimensional curve along a straight line. Such surfaces will actually have only one independent variable if a coordinate system having one axis coincident with the straight line is chosen.

If the two independent variables of the explicit equation of the surface, $z = f(x,y)$ are separable in the sense that

$$z = f(x)f(y)$$

then the surface is "orthogonal". In such a case, the x-dependence may fall in one of the classes of Figure 6 while the y-dependence falls in another. Orthogonal surfaces require fewer operations to evaluate over a grid of the domain of x and y because the defining equation only needs to be evaluated once along the x direction and once along the y direction, with all other points evaluated by simple multiplication of the x and y factors appropriate to each point on the (x,y) plane.

1.6. BASIC CURVE AND SURFACE OPERATIONS

There are many simple operations which can be applied to curves and surfaces in order to change them. Knowledge of these operations enables one to adapt a given curve or surface to a particular need and to thus extend the curves and surfaces given in this reference volume to a larger set of mathematical forms. Only a few of the most common operations are presented here. Of these, two (translation and rotation) are "homomorphic", which means that the form of the curve is preserved and merely its position or orientation in space is changed.

1.6.1. Translation

If the coordinates (x,y,z) of a point are changed to

$$x' = x + a$$

$$y' = y + b$$

$$z' = z + c$$

the curve or surface undergoes a *translation* of amount $(-a, -b, -c)$ along the (x,y,z) axes.

1.6.2. Rotation

In polar coordinates, if the angle θ is changed by

$$\theta' = \theta + \alpha$$

the curve undergoes a *rotation* clockwise of α degrees. This is convenient for polar coordinates, but the rotation can also be expressed in Cartesian coordinates as

$$x' = x \cdot \cos(\alpha) - y \cdot \sin(\alpha)$$

$$y' = x \cdot \sin(\alpha) + y \cdot \cos(\alpha)$$

In three dimensions, a surface can be rotated about any of the three axes by using these equations on the coordinate pairs (x,y), (y,z), or (x,z) depending on whether the rotation is about the z, x, or y axis, respectively.

1.6.3. Linear Scaling

The relations for three-dimensional linear scaling are

$$x' = ax; \quad y' = by; \quad z' = cz$$

These stretch the curve or surface by the factors a, b, and c along the respective axes. When using polar, cylindrical, or spherical coordinates, a similar relation

$$r' = dr$$

stretches or compresses the curve or surface along the radial coordinate by the factor d.

1.6.4. Reflection

A two-dimensional curve is reflected about the x axis by letting

$$y' = -y$$

or about the y axis by letting

$$x' = -x$$

or through the origin by applying both these equations. In three dimensions, a curve or surface is reflected across the y-z, x-z, or x-y planes when

$$x' = -x$$

$$y' = -y$$

$$z' = -z$$

respectively. It can be reflected through the origin when one sets

$$r' = -r$$

in spherical coordinates and mirrored through the z axis when the same operation is made on r for cylindrical coordinates. The application to two-dimensional polar coordinates follows from the cylindrical case.

1.6.5. Rotational Scaling

For two dimensions, let

$$\theta' = c\theta$$

for the polar angle; the polar curve is then "stretched" or "compressed" along the angular direction by a factor c. The same operation can be applied to θ for cylindrical coordinates in three dimensions or to both θ and ϕ for spherical coordinates in three dimensions.

1.6.6. Radial Translation

In two dimensions, if the radial coordinate is translated according to

$$r' = r + a$$

then the entire curve moves outward by the amount a from the origin. Note that this operation is not homomorphic like Cartesian translation because the curve is stretched in the angular direction while undergoing the radial translation. This operation can be performed on the radial coordinate of either cylindrical or spherical coordinate systems in three dimensions.

1.6.7. Weighting

In a two-dimensional Cartesian system, let

$$y' = |x|^a y$$

This operation weights the curve by the factor $|x|^a$, a symmetric operator. If $a > 0$, the curve is stretched in the y direction by a factor which increases with x; but if $a < 0$, the curve is compressed by a factor which decreases with x. Similar treatments can be performed on surfaces in three dimensions.

1.6.8. Nonlinear Scaling

If in two dimensions the scaling

$$y' = y^a$$

is performed, the curve is progressively scaled upward or downward in absolute value, according to whether $a > 1$ or $a < 1$, respectively. Note that, if $y < 0$ and $a = 2,4,6,\ldots$, then the scaled curve will flip to the opposite side of the x axis. Similar scalings can be made in three dimensions using any of the appropriate coordinate systems.

1.6.9. Shear

A curve undergoes "simple shear" when either all its x coordinates or all its y coordinates remain constant while the other set is increased in proportion to x or y, respectively. The general transformations for simple shearing of a two-dimensional curve are

$$x' = x + ay$$

$$y' = bx + y$$

The transformations for simple x shear are

$$x' = x + ay$$

$$y' = y$$

and for simple y shear are

$$x' = x$$

$$y' = y + bx$$

Surfaces may be simply sheared along one or two axes with similar transformations. Another special case of shear is termed "pure shear", and the transformations for a two-dimensional curve are given by

$$x' = kx$$

$$y' = k^{-1}y$$

For surfaces, pure shear will only apply to two of the three coordinate directions, with the remaining one having no change. Pure shear is a special case of linear scaling under this circumstance.

Table 1

Operation	Matrix	Notes
Rotation	$\begin{pmatrix} \cos\theta & \sin\theta \\ -\sin\theta & \cos\theta \end{pmatrix}$	θ is clockwise angle
Linear scaling	$\begin{pmatrix} a & 0 \\ 0 & d \end{pmatrix}$	
Reflection	$\begin{pmatrix} \pm 1 & 0 \\ 0 & \pm 1 \end{pmatrix}$	Use + or − according to whether reflection is about x-axis, y-axis, or origin
Weighting	$\begin{pmatrix} 1 & 0 \\ 0 & x^a \end{pmatrix}$	
Nonlinear scaling	$\begin{pmatrix} 1 & 0 \\ 0 & y^a \end{pmatrix}$	
Simple shear	$\begin{pmatrix} 1 & b \\ c & 1 \end{pmatrix}$	
Rotational scaling	$\begin{pmatrix} 1 & 0 \\ 0 & c \end{pmatrix}$	Use with (r,θ) coordinates

1.6.10. Matrix Method for Transformation

The foregoing transformations can all be expressed in matrix form, which is often convenient for computer algorithms. This is especially true when several transformations are concatenated together, for the matrices can then be simply multiplied together to obtain a single transformation matrix. Given a pair of coordinates (x,y), a matrix transformation to obtain the new coordinates (x′,y′) is written as

$$(x' \;\; y') = (x \;\; y)\begin{pmatrix} a & b \\ c & d \end{pmatrix}$$

or explicitly

$$x' = ax + cy$$

$$y' = bx + dy$$

According to this definition, Table 1 lists several of the x-y transformations discussed previously with their corresponding matrix.

Translations cannot be treated with the above matrix definition. An extension is required to produce what is commonly referred to as the "homogeneous coordinate representation" in computer graphics programming. In its simplest form, an additional coordinate of unity is appended to the (x,y) pair to give (x,y,1). A translation by u and v in the x and y directions is then written using a 3 by 3 matrix

Translations cannot be treated with the above matrix definition. An extension is required to produce what is commonly referred to as the "homogeneous coordinate representation" in computer graphics programing. In its simplest form, an additional coordinate of unity is appended to the (x,y) pair to give (x,y,1). A translation by u and v in the x and y directions is then written using a 3 by 3 matrix

$$(x' \ y' \ 1) = (x \ y \ 1)\begin{pmatrix} 1 & 0 & 0 \\ 0 & 1 & 0 \\ u & v & 1 \end{pmatrix}$$

where, explicitly,

$$x' = x + u$$

$$y' = y + v$$

$$1' = 1$$

With this representation, a radial translation by s units of a curve given in (r,θ) coordinates is effected by

$$(r' \ \theta' \ 1) = (r \ \theta \ 1)\begin{pmatrix} 1 & 0 & 0 \\ 0 & 1 & 0 \\ s & 0 & 1 \end{pmatrix}$$

such that $r' = r + s$ and θ is unchanged.

1.7. METHOD OF PRESENTATION

This reference work is basically intended to be illustrative; therefore all functions, whether curves or surfaces, presented in this volume will have an accompanying plot showing the form of the function. The plot will in almost all cases be on the right-hand page while the equation will be on the facing left-hand page. Curves and surfaces and their plots are numbered for easy reference and grouped according to type. Wherever popular names exist for certain curves or surfaces, they are placed with the equations themselves.

1.7.1. Equations

The equation of each algebraic or transcendental curve will be given in the explicit form $y = f(x)$ or $r = f(\theta)$ wherever possible; similarly, surfaces will be given as $z = f(x,y)$ or $r = f(\theta,z)$ or $r = f(\theta,\phi)$. Whenever polar, cylindrical, or spherical coordinate forms are used, the equation is also written in Cartesian coordinates. Because some curves and surfaces are not amenable to explicit forms, the parametric equations will be used as the alternative when necessary. In either case, whether explicit or parametric, the implicit functional form will also be given, if derivable. The explicit or parametric form is needed in order to evaluate the curve or surface on a computer while the implicit form enables one to determine the degree of the equation (if algebraic) and also easily determine the derivatives in some cases. Notes pertinent to evaluation are given whenever they may help to implement computer algorithms.

For integral curves and surfaces, the equation will be given as the integral $y =$

$\int f(x)$ or $z = \int f(x,y)$. Most of the integral forms have commonly used names (for example, "Bessel functions"). Other curves or surfaces in this reference work are expressed not by single equations but rather by some set of mathematical rules. The method of presentation will vary in these cases, always with the objective of providing the reader with a means of easily constructing the curve or surface by machine computation.

1.7.2. Plots

Plots of two-dimensional curves will be done on the x-y plane, with the x and y axes being horizontal and vertical, respectively. The domain of x and the range of y, unless otherwise stated, will be -1 to $+1$; and the variable form of the curve will be adequately illustrated by a suitable choice of x and y scaling factors and of the constants in the equation. For example, the curve $y = \sin(x)$ can be illustrated for a domain larger than ± 1 by actually plotting $y = \sin(ax)$, with $a > 1$, while still letting x vary between -1 and $+1$. Similarly, the range of y can be limited to ± 1 by plotting $y = c \cdot f(x)$ where the constant c is suitably chosen. Three-dimensional curves and surfaces will have the additional z axis, also from -1 to $+1$, and will be plotted in a projection which satisfactorily illustrates the form of each function. Simple equations will be illustrated by a single plotted curve or surface while more complicated equations may have two or more such plots with different constants in order to indicate the variation possible in a family of curves or surfaces.

In the case of curves which are unbounded in y (for example, $y = 1/x$), a search algorithm has been used to compute and plot the curve point at exactly $y = +1$ or $y = -1$. Curves expressed in polar coordinates (r,θ) are similarly truncated at $r = 1$ in the case that r is unbounded.

REFERENCES

1. **Buck, R. C.,** *Advanced Calculus,* McGraw-Hill, New York, 1965, chap. 5.
2. **Poston, T. and Stewart, I.,** *Catastrophe Theory and Its Applications,* Pitman, New York, 1978.
3. **Abramowitz, M. and Stegun I. A., Eds.,** Handbook of Mathematical Functions, National Bureau of Standards, Department of Commerce, Washington, D.C., 1964.
4. **Mandelbrot, B. B.,** *The Fractal Geometry of Nature,* W. H. Freeman, San Francisco, 1983.
5. **Hayes, B.,** Computer recreations: turning turtle gives one a view of geometry from the inside out, *Sci. Am.,* 250, 14, 1984.

Chapter 2

ALGEBRAIC CURVES

The curves of this chapter are mostly familiar equations found in tables of integrals. Many have acquired traditional or accepted names in the mathematical literature, and these are included wherever appropriate. The last section deals with curves more readily expressed in polar coordinates; this allows much easier computation of the curves than with the form $y = f(x)$, especially when curves are multiple-valued in this form. For curves involving radicals, both the positive and negative branches are plotted to show the symmetry.

2.1. FUNCTIONS WITH $x^{n/m}$

2.1.1. $y = cx^n$ $\qquad$ $y - cx^n = 0$

1. $c = 1, n = 1$ (linear)
2. $c = 1, n = 3$ (cubic)
3. $c = 1, n = 5$ (quintic)
4. $c = 1, n = 2$ (quadratic, or simple parabola)
5. $c = 1, n = 4$ (quartic)
6. $c = 1, n = 6$ (sextic)

2.1.2. $y = c/x^n$ $\qquad$ $yx^n - c = 0$

1. $c = 0.01, n = 1$ (hyperbola)
2. $c = 0.01, n = 3$
3. $c = 0.01, n = 5$
4. $c = 0.01, n = 2$
5. $c = 0.01, n = 4$
6. $c = 0.01, n = 6$

2.1.3. $y = cx^{n/m}$ $\qquad$ $y - cx^{n/m} = 0$

1. $c = 1, n = 1, m = 4$
2. $c = 1, n = 1, m = 2$
3. $c = 1, n = 3, m = 4$
4. $c = 1, n = 5, m = 4$
5. $c = 1, n = 3, m = 2$ (semicubical parabola)
6. $c = 1, n = 7, m = 4$
7. $c = 1, n = 1, m = 3$
8. $c = 1, n = 2, m = 3$
9. $c = 1, n = 4, m = 3$
10. $c = 1, n = 5, m = 3$

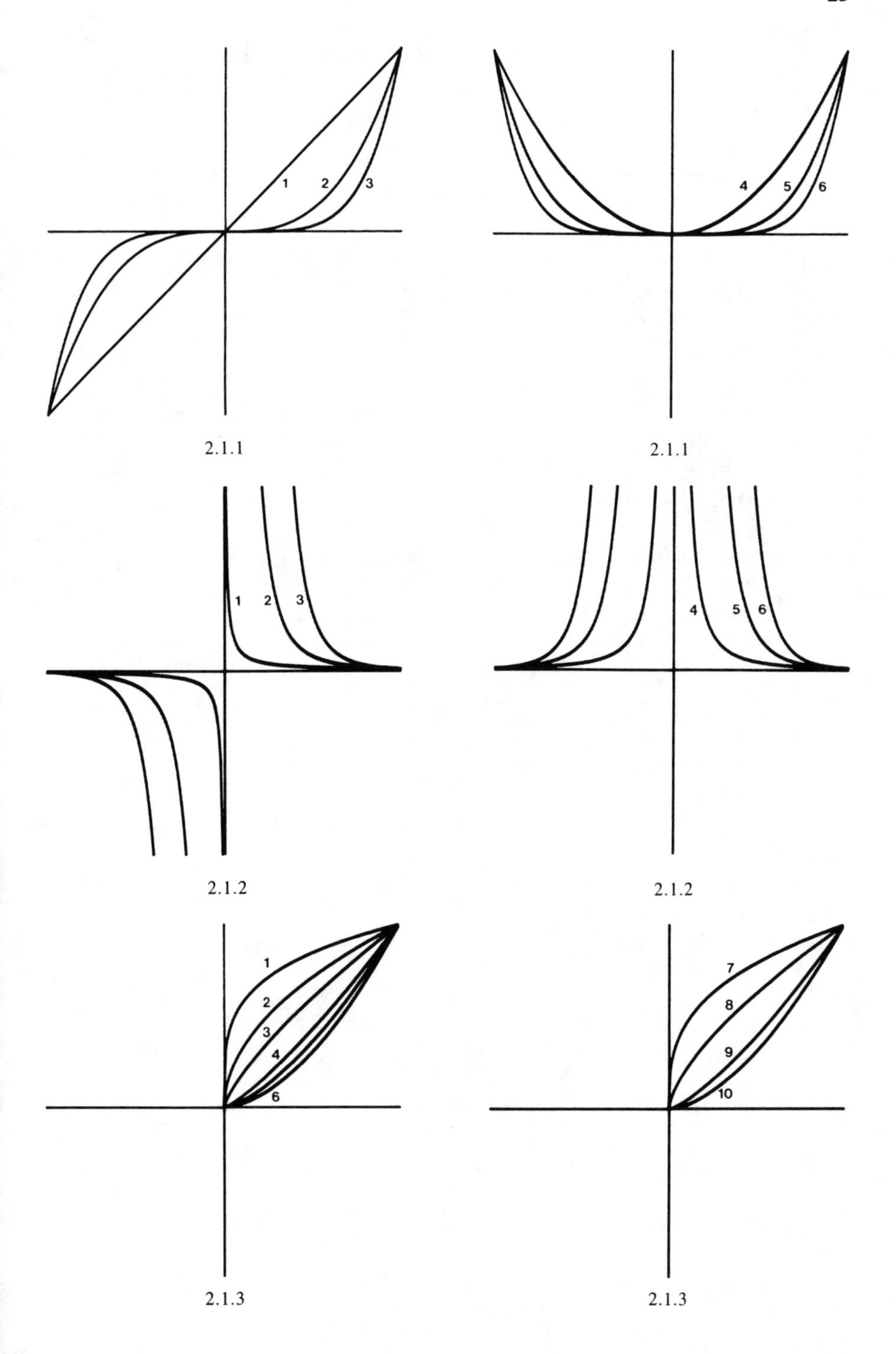

2.1.1 2.1.1

2.1.2 2.1.2

2.1.3 2.1.3

2.2. FUNCTIONS WITH x^n AND $(a + bx)^m$

2.2.1. $y = a + bx$ — $y - bx - a = 0$

1. $a = 0.5$, $b = 0.5$
2. $a = 0.5$, $b = 1.0$
3. $a = 0.5$, $b = 2.0$

2.2.2. $y = (a + bx)^2$ — $y - b^2x^2 - 2abx - a^2 = 0$

1. $a = 0.5$, $b = 0.5$
2. $a = 0.5$, $b = 1.0$
3. $a = 0.5$, $b = 2.0$

2.2.3. $y = (a + bx)^3$ — $y - b^3x^3 - 3ab^2x^2 - 3a^2bx - a^3 = 0$

1. $a = 0.5$, $b = 0.5$
2. $a = 0.5$, $b = 1.0$
3. $a = 0.5$, $b = 2.0$

2.2.4. $y = x(a + bx)$ — $y - bx^2 - ax = 0$

1. $a = 0.5$, $b = 0.5$
2. $a = 0.5$, $b = 1.0$
3. $a = 0.5$, $b = 2.0$

2.2.5. $y = x(a + bx)^2$ — $y - b^2x^3 - 2abx^2 - a^2x = 0$

1. $a = 0.5$, $b = 0.5$
2. $a = 0.5$, $b = 1.0$
3. $a = 0.5$, $b = 2.0$

2.2.6. $y = x(a + bx)^3$ — $y - b^3x^4 - 3ab^2x^3 - 3a^2bx^2 - a^3x = 0$

1. $a = 0.5$, $b = 0.5$
2. $a = 0.5$, $b = 1.0$
3. $a = 0.5$, $b = 2.0$

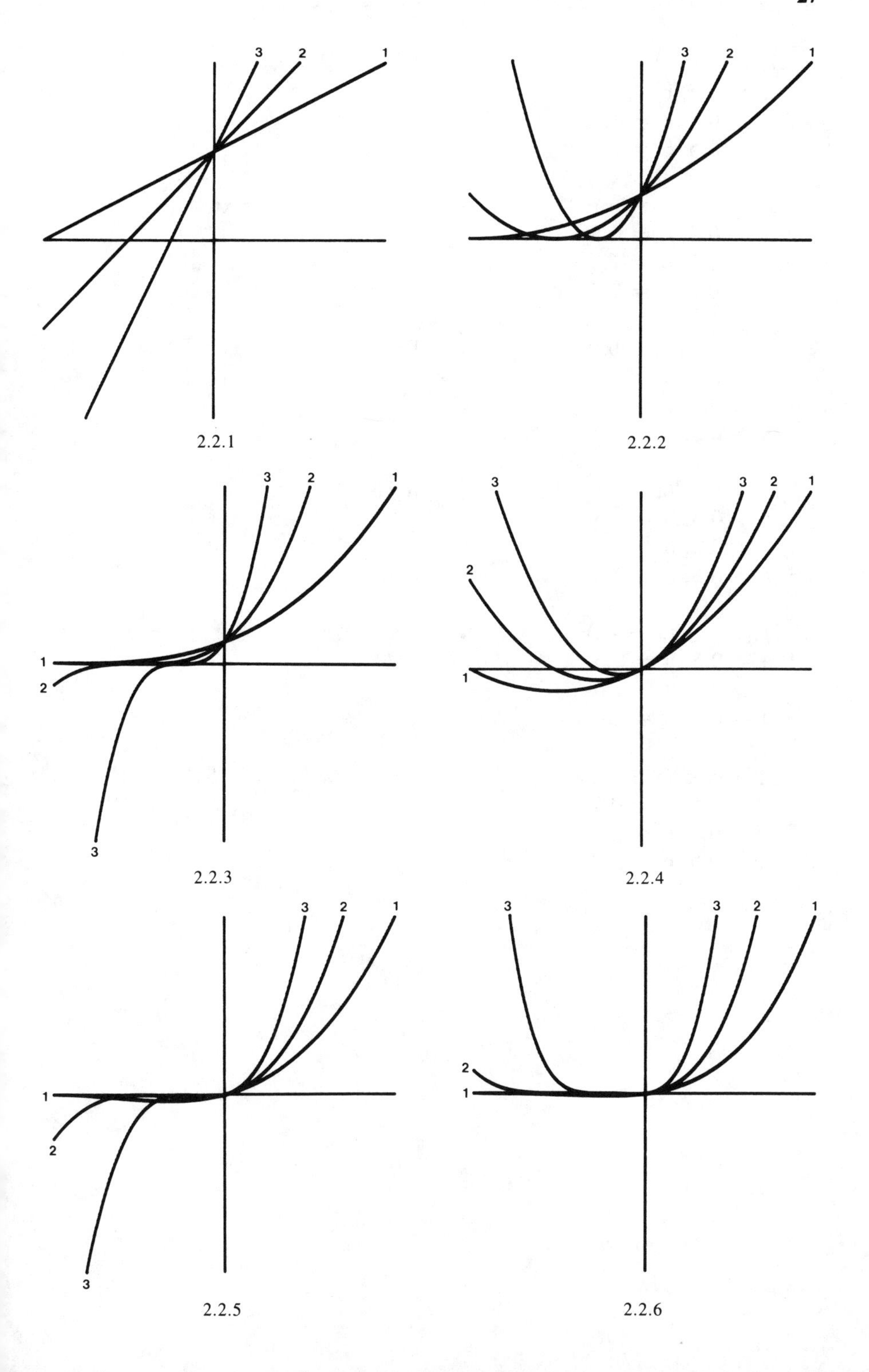

2.2.1

2.2.2

2.2.3

2.2.4

2.2.5

2.2.6

2.2.7. $y = x^2(a + bx)$ $y - bx^3 - ax^2 = 0$
1. $a = 0.5, b = 0.5$
2. $a = 0.5, b = 1.0$
3. $a = 0.5, b = 2.0$

2.2.8. $y = x^2(a + bx)^2$ $y - b^2x^4 - 2abx^3 - a^2x^2 = 0$
1. $a = 0.5, b = 0.5$
2. $a = 0.5, b = 1.0$
3. $a = 0.5, b = 2.0$

2.2.9. $y = x^2(a + bx)^3$ $y - b^3x^5 - 3ab^2x^4 - 3a^2bx^3 - a^3x^2 = 0$
1. $a = 0.5, b = 0.5$
2. $a = 0.5, b = 1.0$
3. $a = 0.5, b = 2.0$

2.2.10. $y = x^3(a + bx)$ $y - bx^4 - ax^3 = 0$
1. $a = 0.5, b = 0.5$
2. $a = 0.5, b = 1.0$
3. $a = 0.5, b = 2.0$

2.2.11. $y = x^3(a + bx)^2$ $y - b^2x^5 - 2abx^4 - a^2x^3 = 0$
1. $a = 0.5, b = 0.5$
2. $a = 0.5, b = 1.0$
3. $a = 0.5, b = 2.0$

2.2.12. $y = x^3(a + bx)^3$ $y - b^3x^6 - 3ab^2x^5 - 3a^2bx^4 - a^3x^3 = 0$
1. $a = 0.5, b = 0.5$
2. $a = 0.5, b = 1.0$
3. $a = 0.5, b = 2.0$

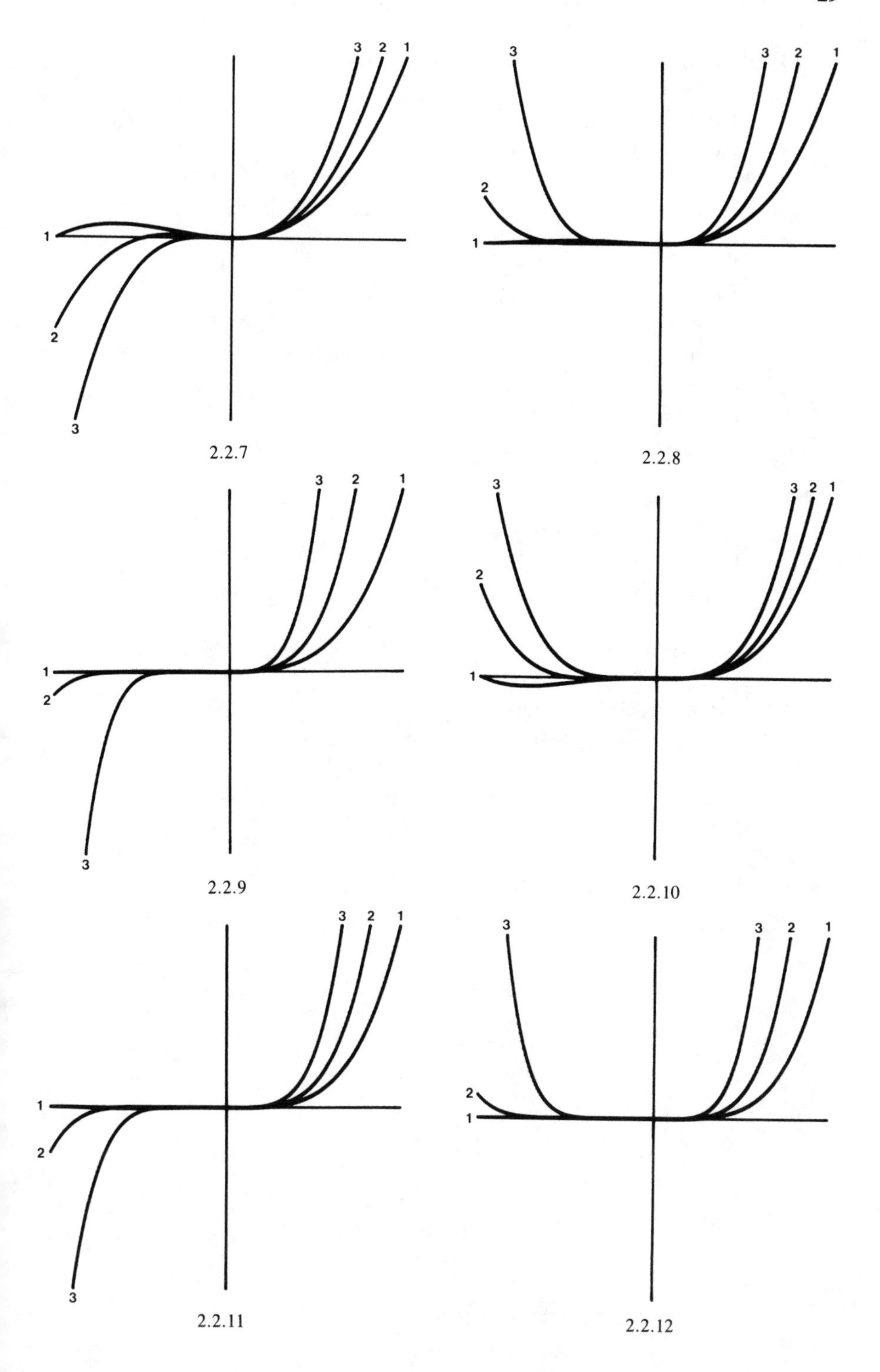
3 2 1
1
2
3
2.2.7
3
3 2 1
2
1
2.2.8
3 2 1
1
2
3
2.2.9
3
3 2 1
2
1
2.2.10
3 2 1
1
2
3
2.2.11
3
3 2 1
2
1
2.2.12

2.2.13. $y = c/(a + bx)$ — $ay + bxy - c = 0$
1. $a = 1.0$, $b = 2.0$, $c = 0.02$
2. $a = 1.0$, $b = 3.0$, $c = 0.02$
3. $a = 1.0$, $b = 4.0$, $c = 0.02$

2.2.14. $y = c/(a + bx)^2$ — $a^2y + 2abxy + b^2x^2y - c = 0$
1. $a = 1.0$, $b = 2.0$, $c = 0.02$
2. $a = 1.0$, $b = 3.0$, $c = 0.02$
3. $a = 1.0$, $b = 4.0$, $c = 0.02$

2.2.15. $y = c/(a + bx)^3$ — $a^3y + 2a^2bxy + 2ab^2x^2y + b^3x^3y - c = 0$
1. $a = 1.0$, $b = 2.0$, $c = 0.02$
2. $a = 1.0$, $b = 3.0$, $c = 0.02$
3. $a = 1.0$, $b = 4.0$, $c = 0.02$

2.2.16. $y = cx/(a + bx)$ — $ay + bxy - cx = 0$
1. $a = 1.0$, $b = 2.0$, $c = 0.1$
2. $a = 1.0$, $b = 3.0$, $c = 0.1$
3. $a = 1.0$, $b = 4.0$, $c = 0.1$

2.2.17. $y = cx/(a + bx)^2$ — $a^2y + 2abxy + b^2x^2y - cx = 0$
1. $a = 1.0$, $b = 2.0$, $c = 0.02$
2. $a = 1.0$, $b = 3.0$, $c = 0.02$
3. $a = 1.0$, $b = 4.0$, $c = 0.02$

2.2.18. $y = cx/(a + bx)^3$ — $a^3y + 3a^2bxy + 3ab^2x^2y + b^3x^3y - cx = 0$
1. $a = 1.0$, $b = 2.0$, $c = 0.01$
2. $a = 1.0$, $b = 3.0$, $c = 0.01$
3. $a = 1.0$, $b = 4.0$, $c = 0.01$

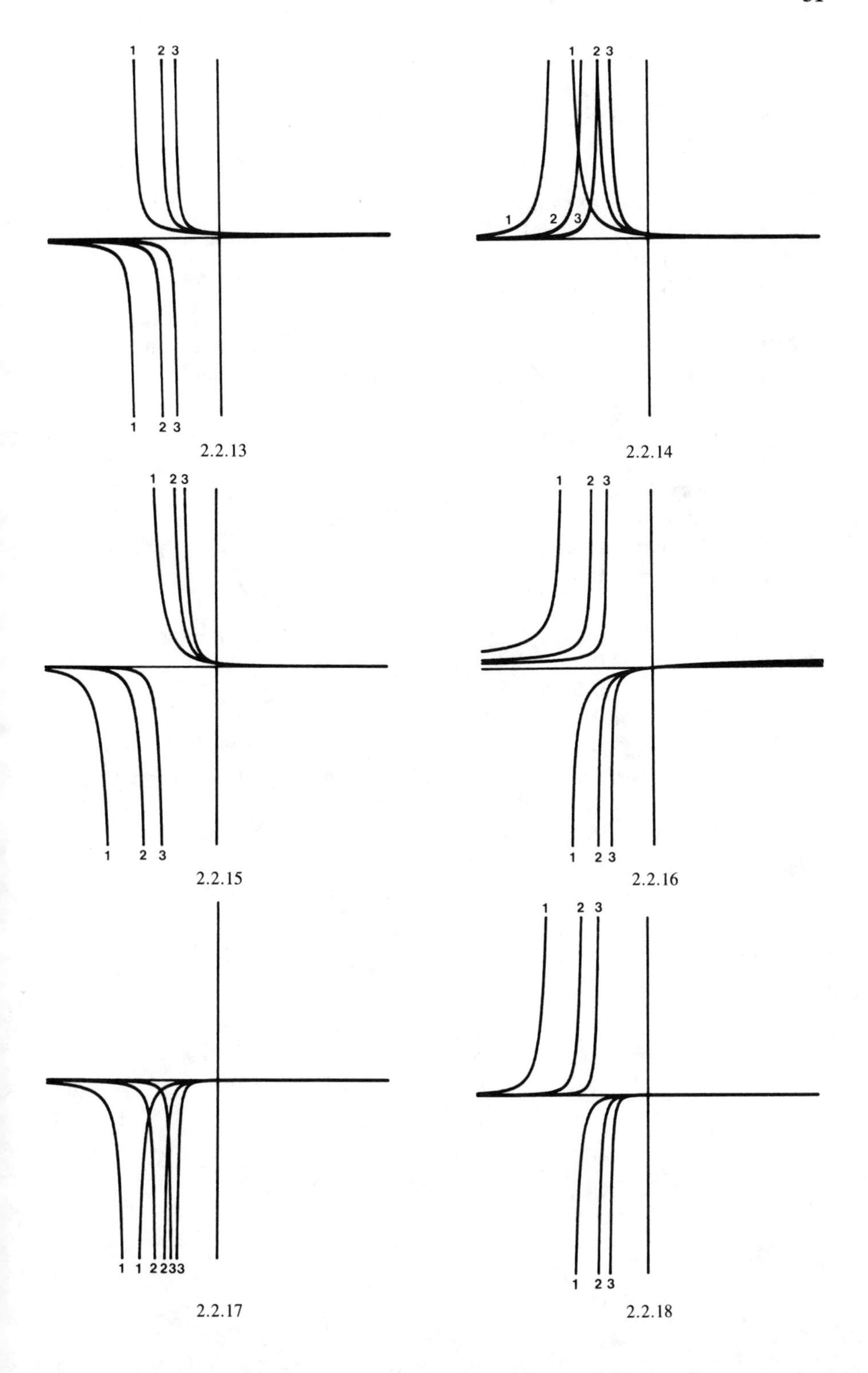

2.2.13

2.2.14

2.2.15

2.2.16

2.2.17

2.2.18

2.2.19. $y = cx^2/(a + bx)$ $ay + bxy - cx^2 = 0$

1. $a = 1.0, b = 2.0, c = 0.2$
2. $a = 1.0, b = 3.0, c = 0.2$
3. $a = 1.0, b = 4.0, c = 0.2$

2.2.20. $y = cx^2/(a + bx)^2$ $a^2y + 2abxy + b^2x^2y - cx^2 = 0$

1. $a = 1.0, b = 2.0, c = 0.1$
2. $a = 1.0, b = 3.0, c = 0.1$
3. $a = 1.0, b = 4.0, c = 0.1$

2.2.21. $y = cx^2/(a + bx)^3$ $a^3y + 3a^2bxy + 3ab^2x^2y + b^3x^3y - cx^2 = 0$

1. $a = 1.0, b = 2.0, c = 0.02$
2. $a = 1.0, b = 3.0, c = 0.02$
3. $a = 1.0, b = 4.0, c = 0.02$

2.2.22. $y = cx^3/(a + bx)$ $ay + bxy - cx^3 = 0$

1. $a = 1.0, b = 2.0, c = 1.0$
2. $a = 1.0, b = 3.0, c = 1.0$
3. $a = 1.0, b = 4.0, c = 1.0$

2.2.23. $y = cx^3/(a + bx)^2$ $a^2y + 2abxy + b^2x^2y - cx^3 = 0$

1. $a = 1.0, b = 2.0, c = 0.2$
2. $a = 1.0, b = 3.0, c = 0.2$
3. $a = 1.0, b = 4.0, c = 0.2$

2.2.24. $y = cx^3/(a + bx)^3$ $a^3y + 3a^2bxy + 3ab^2x^2y + b^3x^3y - cx^3 = 0$

1. $a = 1.0, b = 2.0, c = 0.1$
2. $a = 1.0, b = 3.0, c = 0.1$
3. $a = 1.0, b = 4.0, c = 0.1$

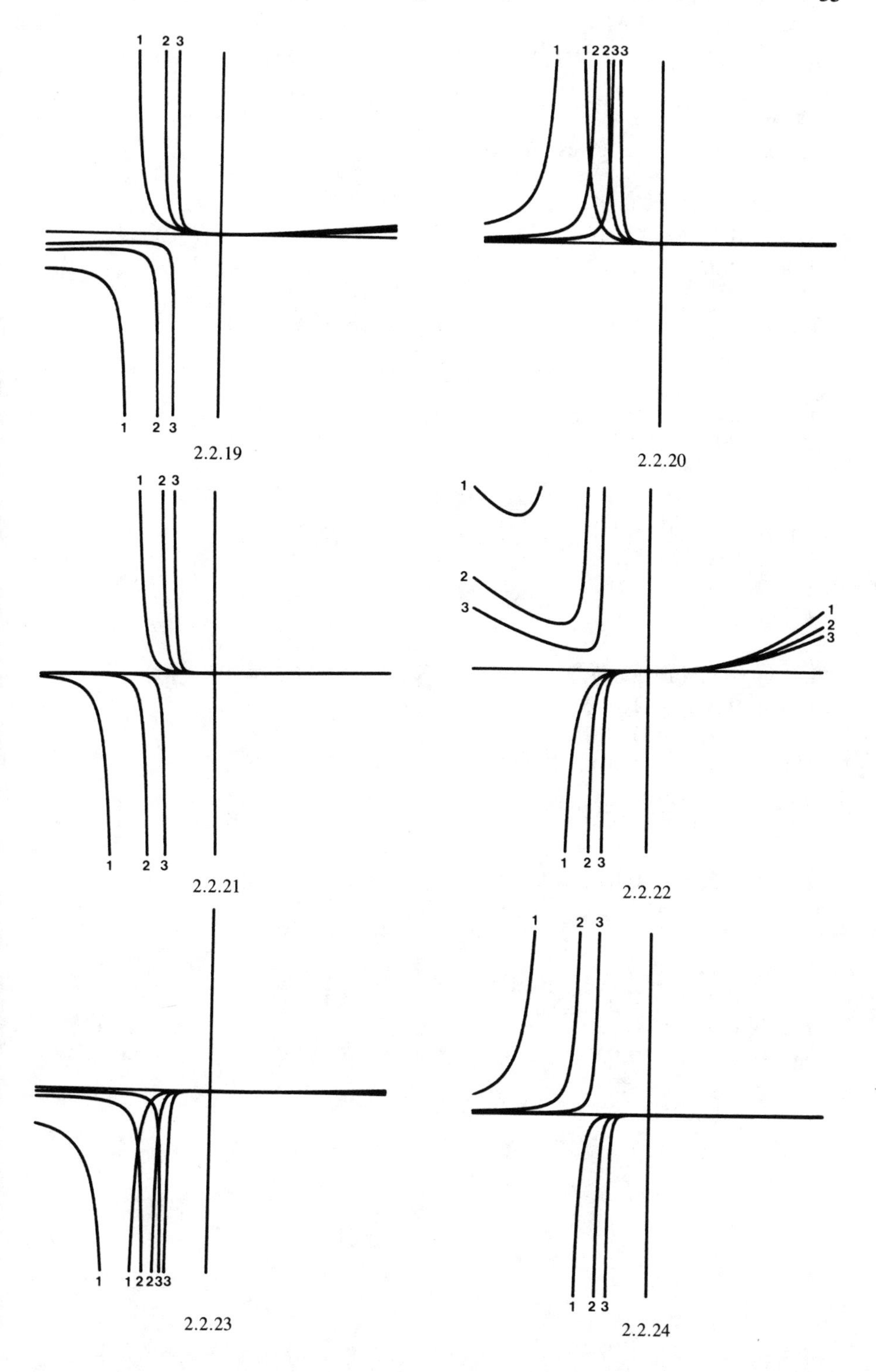

2.2.19

2.2.20

2.2.21

2.2.22

2.2.23

2.2.24

2.3. FUNCTIONS WITH $a^2 + x^2$ AND x^m

2.3.1. $y = c/(a^2 + x^2)$ $a^2y + x^2y - c = 0$

Special case: $c = a^3$ gives "Witch of Agnesi"

1. $a = 0.2$, $c = 0.04$
2. $a = 0.5$, $c = 0.04$
3. $a = 0.8$, $c = 0.04$

2.3.2. $y = cx/(a^2 + x^2)$ $a^2y + x^2y - cx = 0$

"Serpentine"

1. $a = 0.2$, $c = 0.3$
2. $a = 0.5$, $c = 0.3$
3. $a = 0.8$, $c = 0.3$

2.3.3. $y = cx^2/(a^2 + x^2)$ $a^2y + x^2y - cx^2 = 0$

1. $a = 0.2$, $c = 1.0$
2. $a = 0.5$, $c = 1.0$
3. $a = 0.8$, $c = 1.0$

2.3.4. $y = cx^3/(a^2 + x^2)$ $a^2y + x^2y - cx^3 = 0$

1. $a = 0.2$, $c = 1.0$
2. $a = 0.5$, $c = 1.0$
3. $a = 0.8$, $c = 1.0$

2.3.5. $y = c/(x(a^2 + x^2))$ $a^2xy + x^3y - c = 0$

1. $a = 0.2$, $c = 0.02$
2. $a = 0.5$, $c = 0.02$
3. $a = 0.8$, $c = 0.02$

2.3.6. $y = c/(x^2(a^2 + x^2))$ $a^2x^2y + x^4y - c = 0$

1. $a = 0.2$, $c = 0.02$
2. $a = 0.5$, $c = 0.02$
3. $a = 0.8$, $c = 0.02$

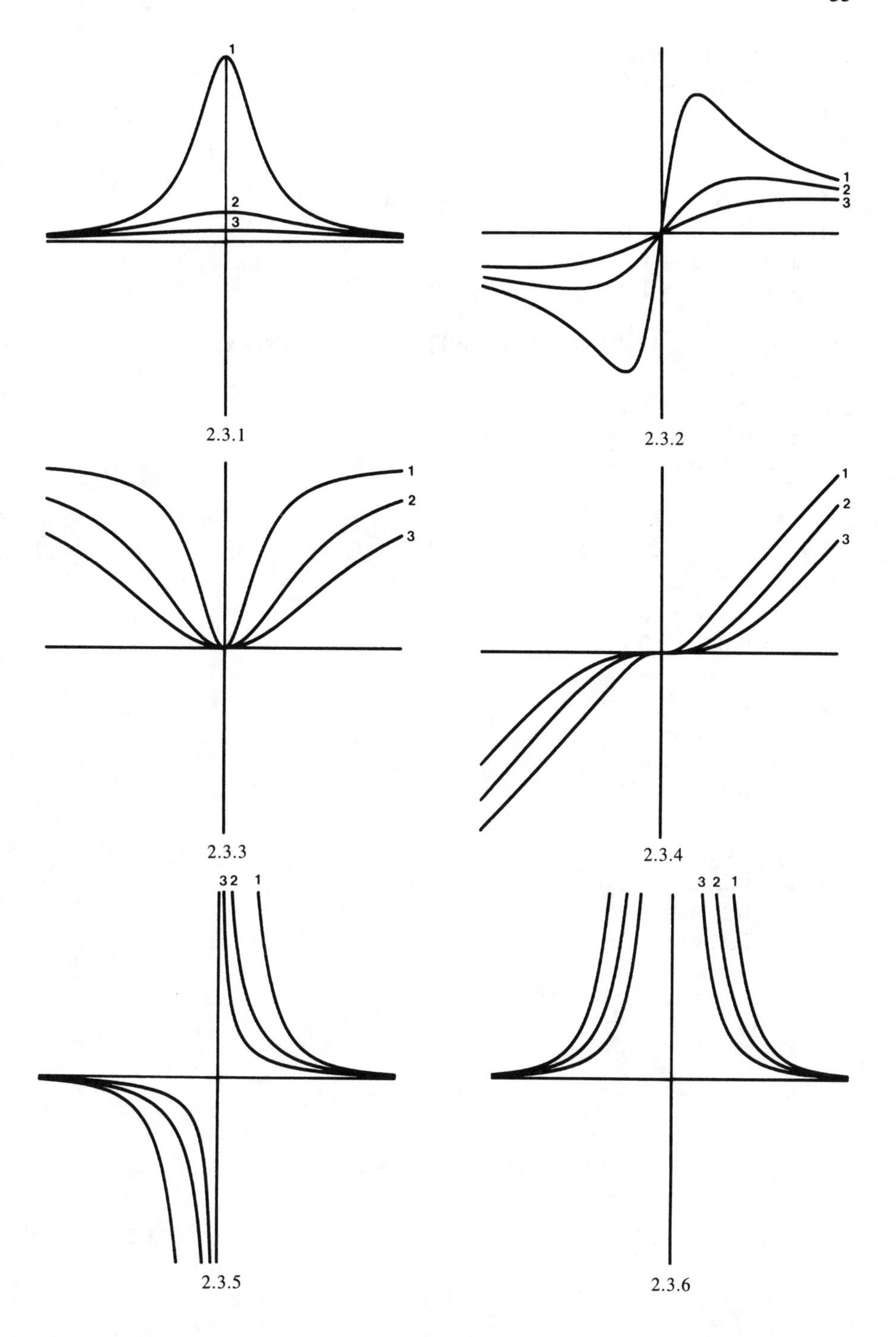

2.3.1

2.3.2

2.3.3

2.3.4

2.3.5

2.3.6

2.3.7. $y = cx(a^2 + x^2)$ $\qquad y - a^2cx - cx^3 = 0$

1. $a = 0.2, c = 1.0$
2. $a = 0.5, c = 1.0$
3. $a = 0.8, c = 1.0$

2.3.8. $y = cx^2(a^2 + x^2)$ $\qquad y - a^2cx^2 - cx^4 = 0$

1. $a = 0.2, c = 1.0$
2. $a = 0.5, c = 1.0$
3. $a = 0.8, c = 1.0$

2.4. FUNCTIONS WITH $a^2 - x^2$ AND x^m

2.4.1. $y = c/(a^2 - x^2)$ $\qquad a^2y - x^2y - c = 0$

1. $a = 0.2, c = 0.03$
2. $a = 0.5, c = 0.03$
3. $a = 0.8, c = 0.03$

2.4.2. $y = cx/(a^2 - x^2)$ $\qquad a^2y - x^2y - cx = 0$

1. $a = 0.2, c = 0.1$
2. $a = 0.5, c = 0.1$
3. $a = 0.8, c = 0.1$

2.4.3. $y = cx^2/(a^2 - x^2)$ $\qquad a^2y - x^2y - cx^2 = 0$

1. $a = 0.2, c = 0.2$
2. $a = 0.5, c = 0.2$
3. $a = 0.8, c = 0.2$

2.4.4. $y = cx^3/(a^2 - x^2)$ $\qquad a^2y - x^2y - cx^3 = 0$

1. $a = 0.2, c = 0.2$
2. $a = 0.5, c = 0.2$
3. $a = 0.8, c = 0.2$

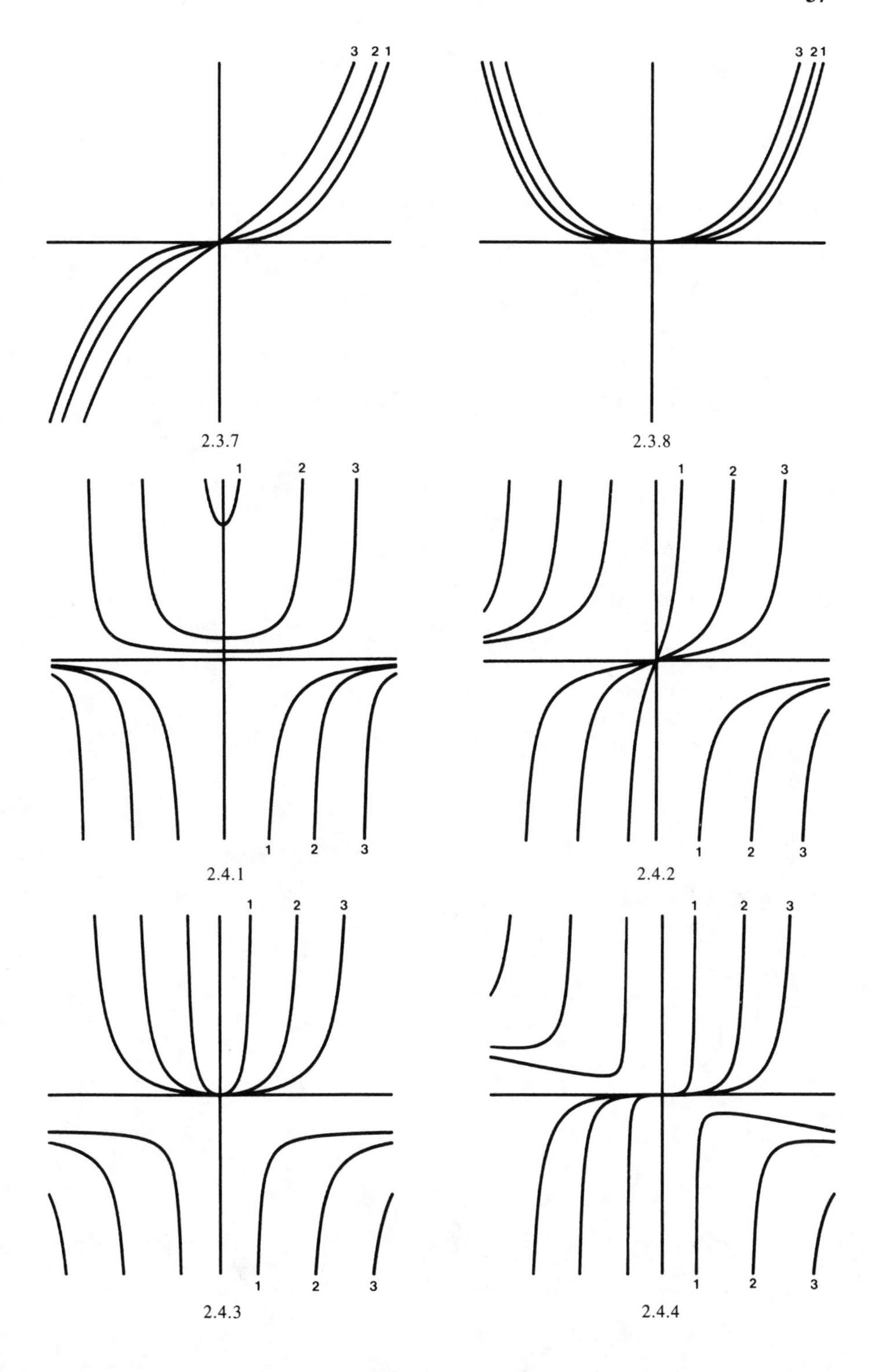

2.3.7

2.3.8

2.4.1

2.4.2

2.4.3

2.4.4

2.4.5. $y = c/(x(a^2 - x^2))$ $a^2xy - x^3y - c = 0$
1. $a = 0.2, c = 0.001$
2. $a = 0.5, c = 0.001$
3. $a = 0.8, c = 0.001$

2.4.6. $y = c/(x^2(a^2 - x^2))$ $a^2x^2y - x^4y - c = 0$
1. $a = 0.2, c = 0.001$
2. $a = 0.5, c = 0.001$
3. $a = 0.8, c = 0.001$

2.4.7. $y = cx(a^2 - x^2)$ $y - a^2cx + cx^3 = 0$
1. $a = 0.2, c = 1.0$
2. $a = 0.5, c = 1.0$
3. $a = 0.8, c = 1.0$

2.4.8. $y = cx^2(a^2 - x^2)$ $y - a^2cx^2 + cx^4 = 0$
1. $a = 0.2, c = 1.0$
2. $a = 0.5, c = 1.0$
3. $a = 0.8, c = 1.0$

2.5. FUNCTIONS WITH $a^3 + x^3$ AND x^m

2.5.1. $y = c/(a^3 + x^3)$ $a^3y + x^3y - c = 0$
1. $a = 0.2, c = 0.01$
2. $a = 0.3, c = 0.01$
3. $a = 0.4, c = 0.01$

2.5.2. $y = cx/(a^3 + x^3)$ $a^3y + x^3y - cx = 0$
1. $a = 0.1, c = 0.01$
2. $a = 0.3, c = 0.01$
3. $a = 0.5, c = 0.01$

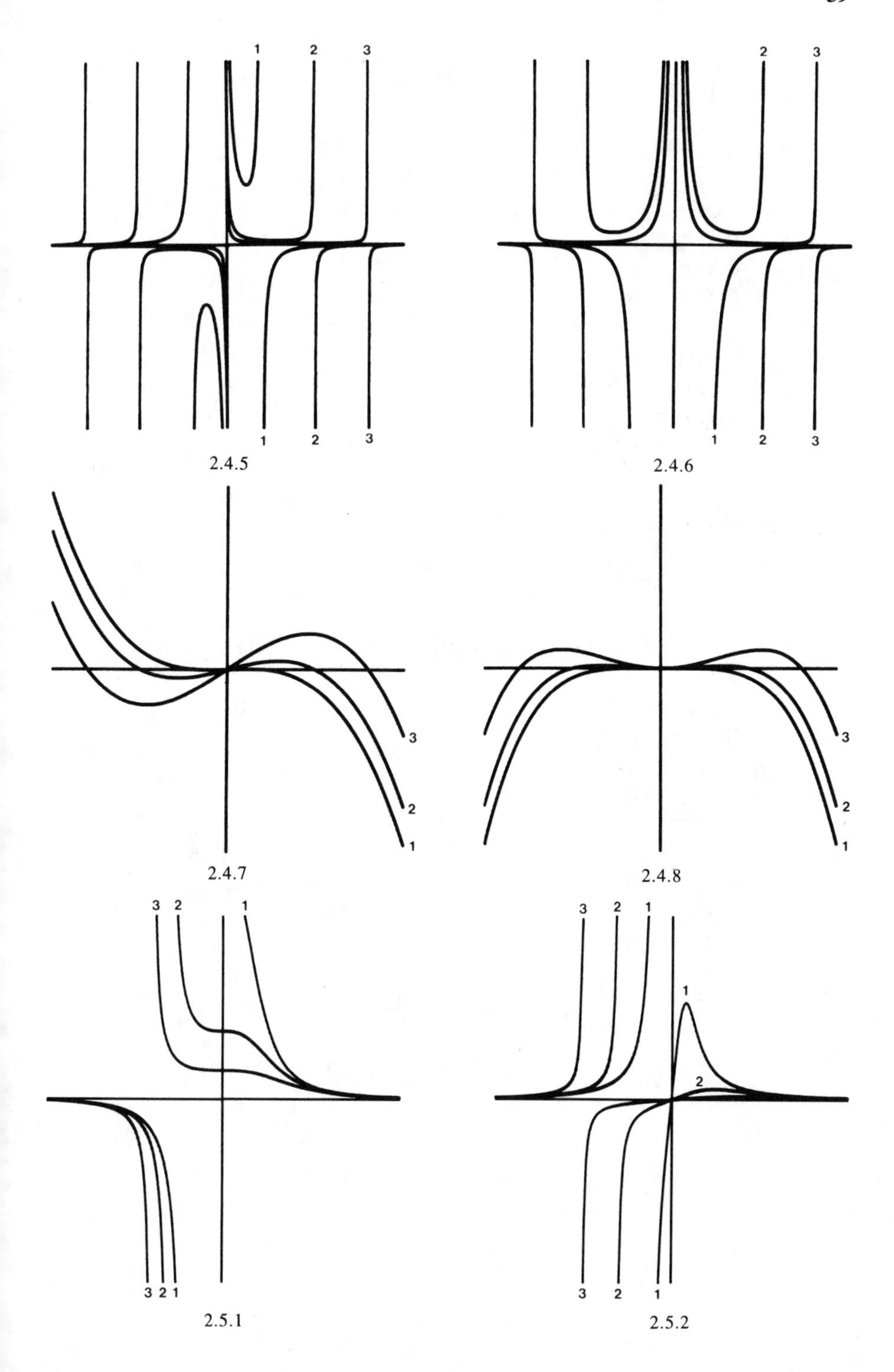
1
2
3
1
2
3
2.4.5
2
3
1
2
3
2.4.6
3
2
1
2.4.7
3
2
1
2.4.8
3
2
1
3
2
1
2.5.1
3
2
1
1
2
3
2
1
2.5.2

2.5.3. $y = cx^2/(a^3 + x^3)$ $a^3y + x^3y - cx^2 = 0$
1. $a = 0.1, c = 0.1$
2. $a = 0.3, c = 0.1$
3. $a = 0.5, c = 0.1$

2.5.4. $y = cx^3/(a^3 + x^3)$ $a^3y + x^3y - cx^3 = 0$
1. $a = 0.1, c = 0.2$
2. $a = 0.3, c = 0.2$
3. $a = 0.5, c = 0.2$

2.5.5. $y = c/(x(a^3 + x^3))$ $a^3xy + x^4y - c = 0$
1. $a = 0.5, c = 0.01$
2. $a = 0.7, c = 0.01$
3. $a = 0.9, c = 0.01$

2.5.6. $y = cx(a^3 + x^3)$ $y - a^3cx - cx^4 = 0$
1. $a = 0.5, c = 0.5$
2. $a = 0.7, c = 0.5$
3. $a = 0.9, c = 0.5$

2.6. FUNCTIONS WITH $a^3 - x^3$ AND x^m

2.6.1. $y = c/(a^3 - x^3)$ $a^3y - x^3y - c = 0$
1. $a = 0.2, c = 0.01$
2. $a = 0.3, c = 0.01$
3. $a = 0.4, c = 0.01$

2.6.2. $y = cx/(a^3 - x^3)$ $a^3y - x^3y - cx = 0$
1. $a = 0.1, c = 0.01$
2. $a = 0.3, c = 0.01$
3. $a = 0.5, c = 0.01$

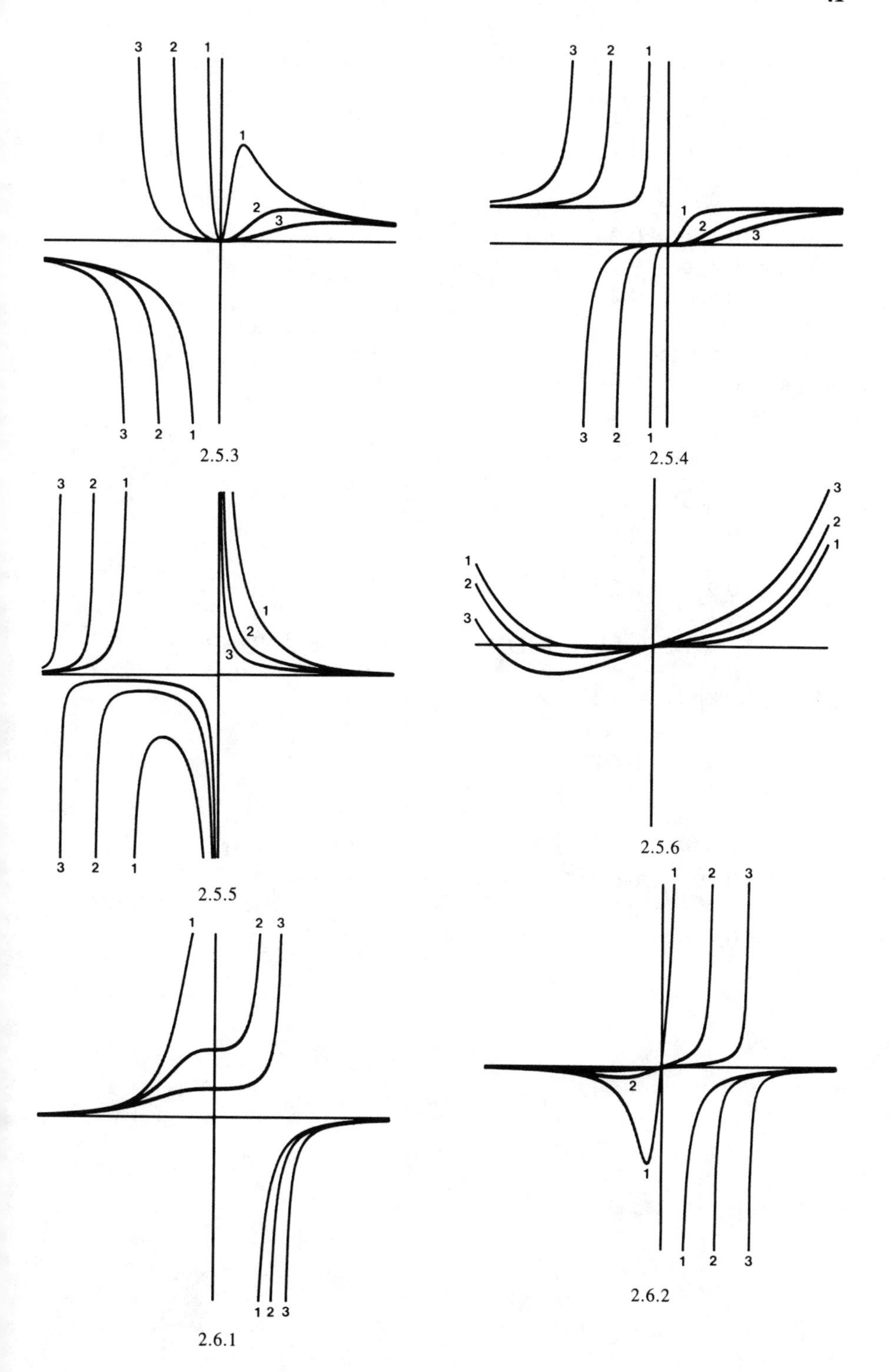
3
2
1
1
2
3
3
2
1
2.5.3
3
2
1
1
2
3
3
2
1
2.5.4
3
2
1
1
2
3
3
2
1
2.5.5
3
2
1
1
2
3
2.5.6
1
2
3
1
2 3
1 2 3
2.6.1
1
2
3
2
1
1
2
3
2.6.2

2.6.3. $y = cx^2/(a^3 - x^3)$ $a^3y - x^3y - cx^2 = 0$

1. $a = 0.1, c = 0.1$
2. $a = 0.3, c = 0.1$
3. $a = 0.5, c = 0.1$

2.6.4. $y = cx^3/(a^3 - x^3)$ $a^3y - x^3y - cx^3 = 0$

1. $a = 0.1, c = 0.2$
2. $a = 0.3, c = 0.2$
3. $a = 0.5, c = 0.2$

2.6.5. $y = c/(x(a^3 - x^3))$ $a^3xy - x^4y - c = 0$

1. $a = 0.5, c = 0.01$
2. $a = 0.7, c = 0.01$
3. $a = 0.9, c = 0.01$

2.6.6. $y = cx(a^3 - x^3)$ $y - a^3cx + cx^4 = 0$

1. $a = 0.5, c = 0.5$
2. $a = 0.7, c = 0.5$
3. $a = 0.9, c = 0.5$

2.7. FUNCTIONS WITH $a^4 + x^4$ AND x^m

2.7.1. $y = c/(a^4 + x^4)$ $a^4y + x^4y - c = 0$

1. $a = 0.3, c = 0.007$
2. $a = 0.4, c = 0.007$
3. $a = 0.5, c = 0.007$

2.7.2. $y = cx/(a^4 + x^4)$ $a^4y + x^4y - cx = 0$

1. $a = 0.2, c = 0.01$
2. $a = 0.3, c = 0.01$
3. $a = 0.4, c = 0.01$

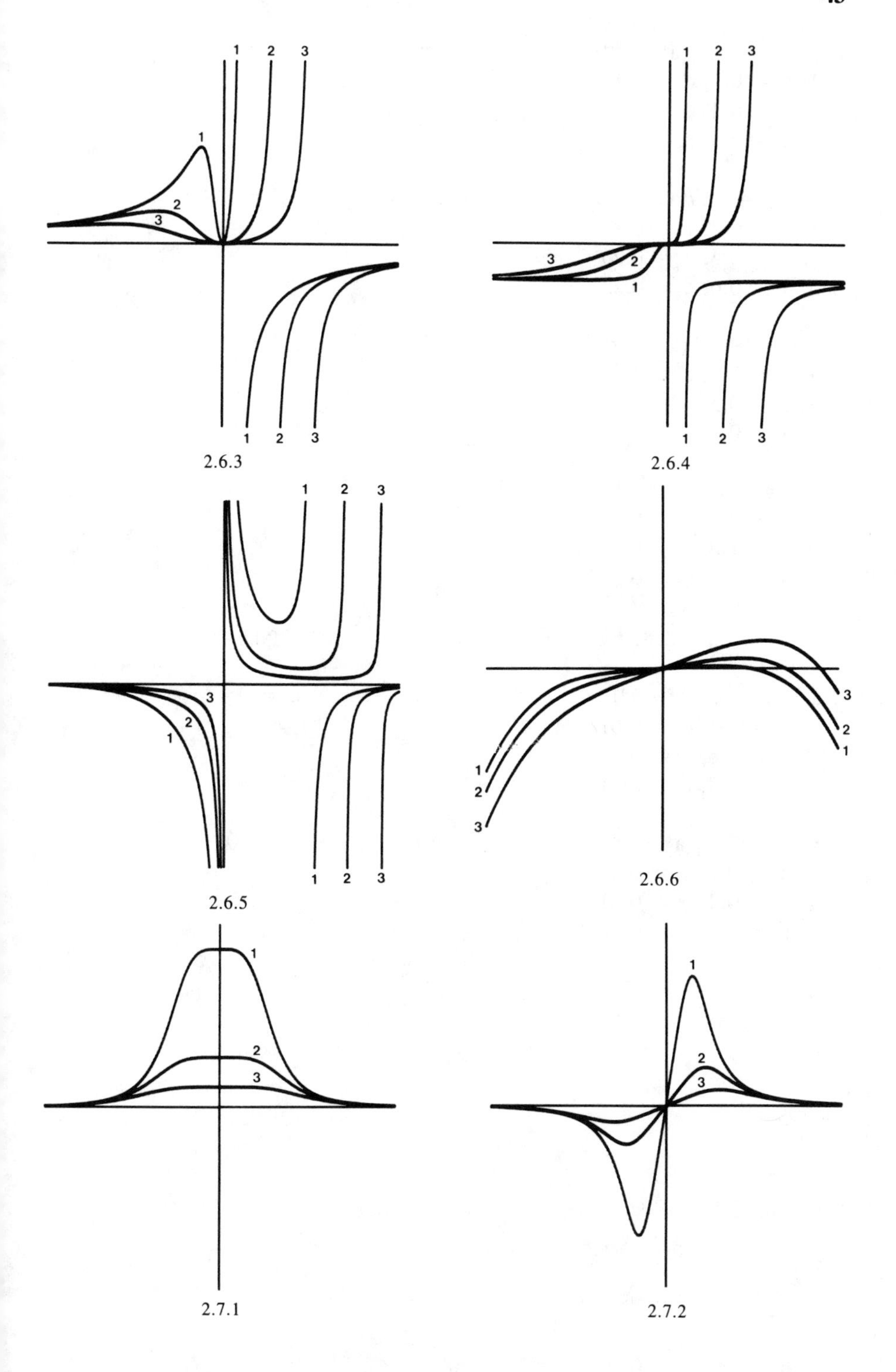

2.6.3

2.6.4

2.6.5

2.6.6

2.7.1

2.7.2

2.7.3. $y = cx^2/(a^4 + x^4)$ $a^4y + x^4y - cx^2 = 0$
1. $a = 0.3, c = 0.15$
2. $a = 0.4, c = 0.15$
3. $a = 0.5, c = 0.15$

2.7.4. $y = cx^3/(a^4 + x^4)$ $a^4y + x^4y - cx^3 = 0$
1. $a = 0.2, c = 0.25$
2. $a = 0.4, c = 0.25$
3. $a = 0.6, c = 0.25$

2.7.5. $y = cx^4/(a^4 + x^4)$ $a^4y + x^4y - cx^4 = 0$
1. $a = 0.2, c = 1.0$
2. $a = 0.5, c = 1.0$
3. $a = 0.8, c = 1.0$

2.7.6. $y = cx(a^4 + x^4)$ $y - a^4cx - cx^5 = 0$
1. $a = 0.5, c = 0.5$
2. $a = 1.0, c = 0.5$
3. $a = 1.2, c = 0.5$

2.8. FUNCTIONS WITH $a^4 - x^4$ AND x^m

2.8.1. $y = c/(a^4 - x^4)$ $a^4y - x^4y - c = 0$
1. $a = 0.4, c = 0.01$
2. $a = 0.6, c = 0.01$
3. $a = 0.8, c = 0.01$

2.8.2. $y = cx/(a^4 - x^4)$ $a^4y - x^4y - cx = 0$
1. $a = 0.2, c = 0.01$
2. $a = 0.4, c = 0.01$
3. $a = 0.6, c = 0.01$

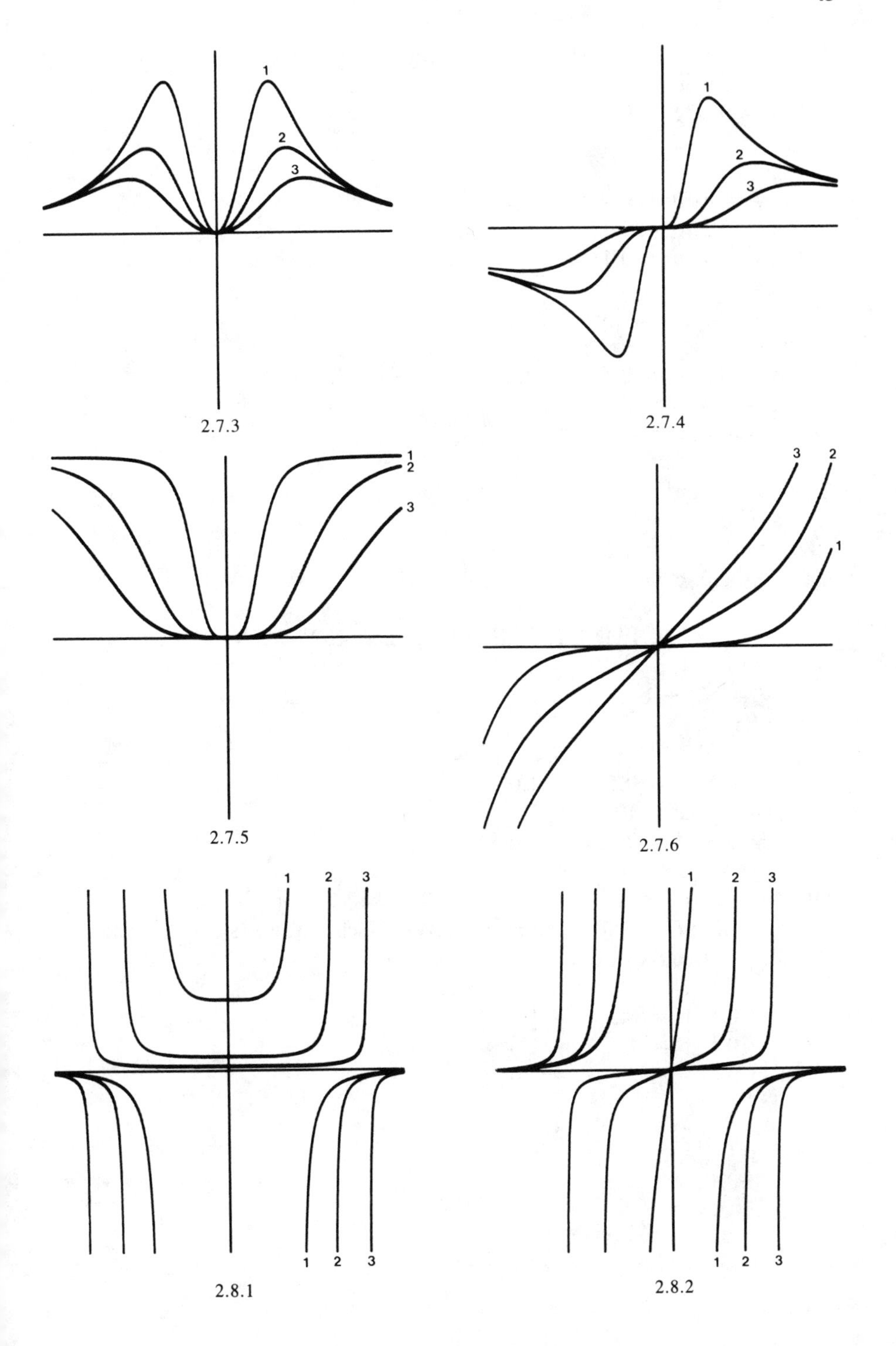

2.7.3 2.7.4 2.7.5 2.7.6 2.8.1 2.8.2

2.8.3. $y = cx^2/(a^4 - x^4)$ $a^4y - x^4y - cx^2 = 0$

1. $a = 0.2$, $c = 0.1$
2. $a = 0.4$, $c = 0.1$
3. $a = 0.6$, $c = 0.1$

2.8.4. $y = cx^3/(a^4 - x^4)$ $a^4y - x^4y - cx^3 = 0$

1. $a = 0.2$, $c = 0.1$
2. $a = 0.4$, $c = 0.1$
3. $a = 0.6$, $c = 0.1$

2.8.5. $y = cx^4/(a^4 - x^4)$ $a^4y - x^4y - cx^4 = 0$

1. $a = 0.2$, $c = 0.1$
2. $a = 0.5$, $c = 0.1$
3. $a = 0.8$, $c = 0.1$

2.8.6. $y = cx(a^4 - x^4)$ $y - a^4cx + cx^5 = 0$

1. $a = 0.4$, $c = 1.0$
2. $a = 0.8$, $c = 1.0$
3. $a = 1.0$, $c = 1.0$

2.9. FUNCTIONS WITH $(a + bx)^{1/2}$ AND x^m

2.9.1. $y = c(a + bx)^{1/2}$ $y^2 - bc^2x - ac^2 = 0$

"Parabola"

1. $a = 0.5$, $b = 0.5$, $c = 1.0$
2. $a = 0.5$, $b = 1.0$, $c = 1.0$
3. $a = 0.5$, $b = 2.0$, $c = 1.0$

2.9.2. $y = cx(a + bx)^{1/2}$ $y^2 - bc^2x^3 - ac^2x^2 = 0$

Special case: $c = 1/(3a)$ and $b = 1$ gives "Tschirnhauser's cubic" (also called "trisectrix of Catalan")

1. $a = 0.5$, $b = 0.5$, $c = 1.0$
2. $a = 0.5$, $b = 1.0$, $c = 1.0$
3. $a = 0.5$, $b = 2.0$, $c = 1.0$

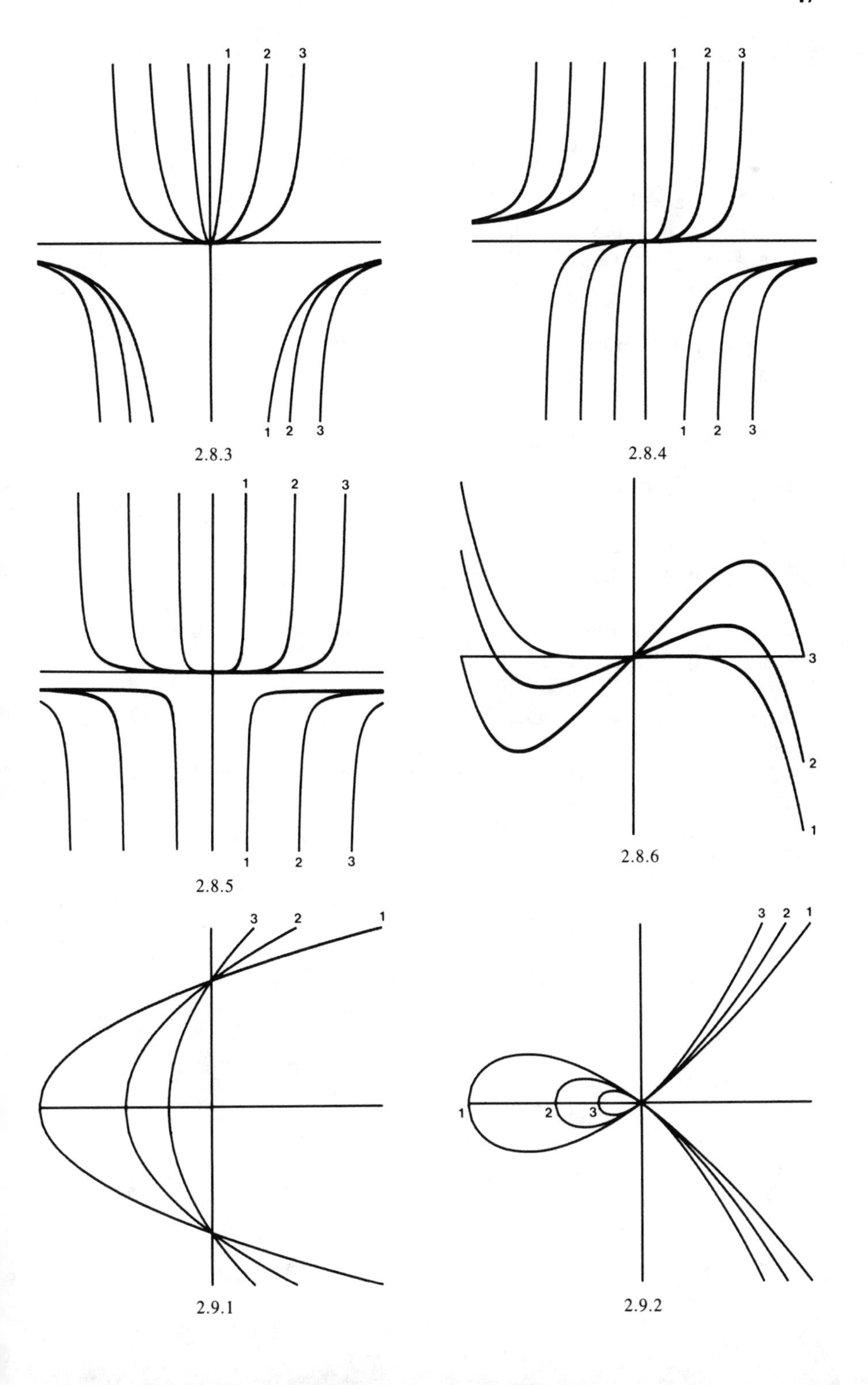
1
2
3
1
2
3
2.8.3
1
2
3
1
2
3
2.8.4
1
2
3
1
2
3
2.8.5
3
2
1
2.8.6
3
2
1
2.9.1
3
2
1
1
2
3
2.9.2

2.9.3. $y = cx^2(a + bx)^{1/2}$ $y^2 - bc^2x^5 - ac^2x^4 = 0$

1. $a = 0.5, b = 0.5, c = 1.0$
2. $a = 0.5, b = 1.0, c = 1.0$
3. $a = 0.5, b = 2.0, c = 1.0$

2.9.4. $y = c(a + bx)^{1/2}/x$ $x^2y^2 - c^2bx - c^2a = 0$

1. $a = 0.5, b = 0.5, c = 0.2$
2. $a = 0.5, b = 1.0, c = 0.2$
3. $a = 0.5, b = 2.0, c = 0.2$

2.9.5. $y = c(a + bx)^{1/2}/x^2$ $x^4y^2 - c^2bx - c^2a = 0$

1. $a = 0.5, b = 0.5, c = 0.1$
2. $a = 0.5, b = 1.0, c = 0.1$
3. $a = 0.5, b = 2.0, c = 0.1$

2.9.6. $y = c/(a + bx)^{1/2}$ $ay^2 + bxy^2 - c^2 = 0$

1. $a = 1.0, b = 1.0, c = 0.5$
2. $a = 1.0, b = 2.0, c = 0.5$
3. $a = 1.0, b = 4.0, c = 0.5$

2.9.7. $y = cx/(a + bx)^{1/2}$ $ay^2 + bxy^2 - c^2x^2 = 0$

1. $a = 1.0, b = 1.0, c = 1.0$
2. $a = 1.0, b = 2.0, c = 1.0$
3. $a = 1.0, b = 4.0, c = 1.0$

2.9.8. $y = cx^2/(a + bx)^{1/2}$ $ay^2 + bxy^2 - c^2x^4 = 0$

1. $a = 1.0, b = 1.0, c = 1.0$
2. $a = 1.0, b = 2.0, c = 1.0$
3. $a = 1.0, b = 4.0, c = 1.0$

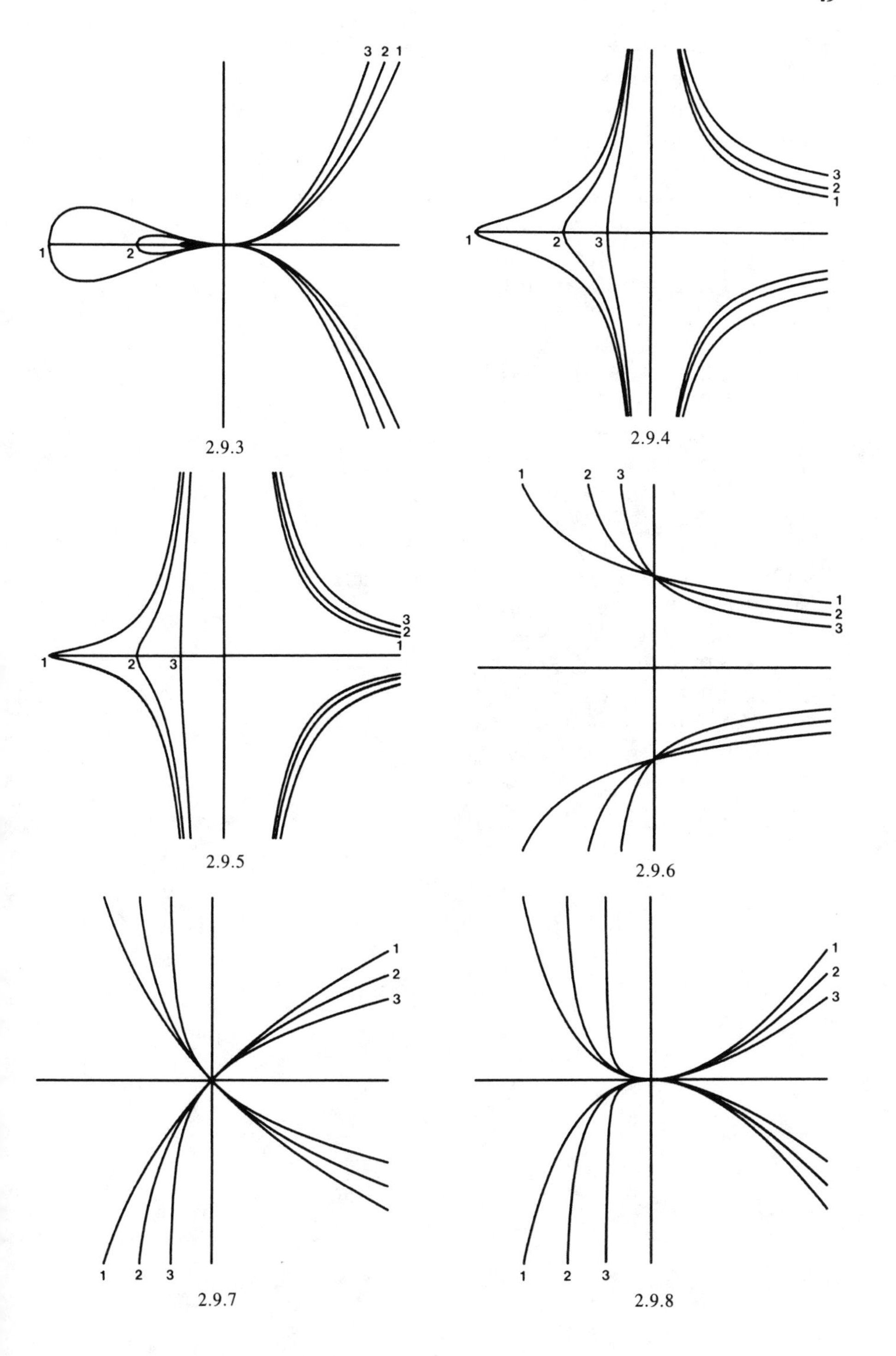

2.9.3

2.9.4

2.9.5

2.9.6

2.9.7

2.9.8

2.9.9. $y = c/(x(a + bx)^{1/2})$ $ax^2y^2 + bx^3y^2 - c^2 = 0$

1. $a = 1.0, b = 0.8, c = 0.2$
2. $a = 1.0, b = 1.0, c = 0.2$
3. $a = 1.0, b = 1.2, c = 0.2$

2.9.10. $y = c/(x^2(a + bx)^{1/2})$ $ax^4y^2 + bx^5y^2 - c^2 = 0$

1. $a = 1.0, b = 0.8, c = 0.1$
2. $a = 1.0, b = 1.0, c = 0.1$
3. $a = 1.0, b = 1.2, c = 0.1$

2.9.11. $y = cx^{1/2}(a + bx)^{1/2}$ $y^2 - ac^2x - bc^2x^2 = 0$

1. $a = 2.0, b = -2.0, c = 1.0$
2. $a = 2.0, b = -3.0, c = 1.0$
3. $a = 2.0, b = -4.0, c = 1.0$
4. $a = 0.3, b = 1.0, c = 1.0$
5. $a = 0.5, b = 1.0, c = 1.0$
6. $a = 0.7, b = 1.0, c = 1.0$

2.9.12. $y = cx^{3/2}(a + bx)^{1/2}$ $y^2 - ac^2x^3 - bc^2x^4 = 0$

Special case: $b < 0$ gives "piriform"

1. $a = 4.0, b = -4.0, c = 1.0$
2. $a = 4.0, b = -6.0, c = 1.0$
3. $a = 4.0, b = -12.0, c = 1.0$
4. $a = 0.3, b = 1.0, c = 1.0$
5. $a = 0.5, b = 1.0, c = 1.0$
6. $a = 0.7, b = 1.0, c = 1.0$

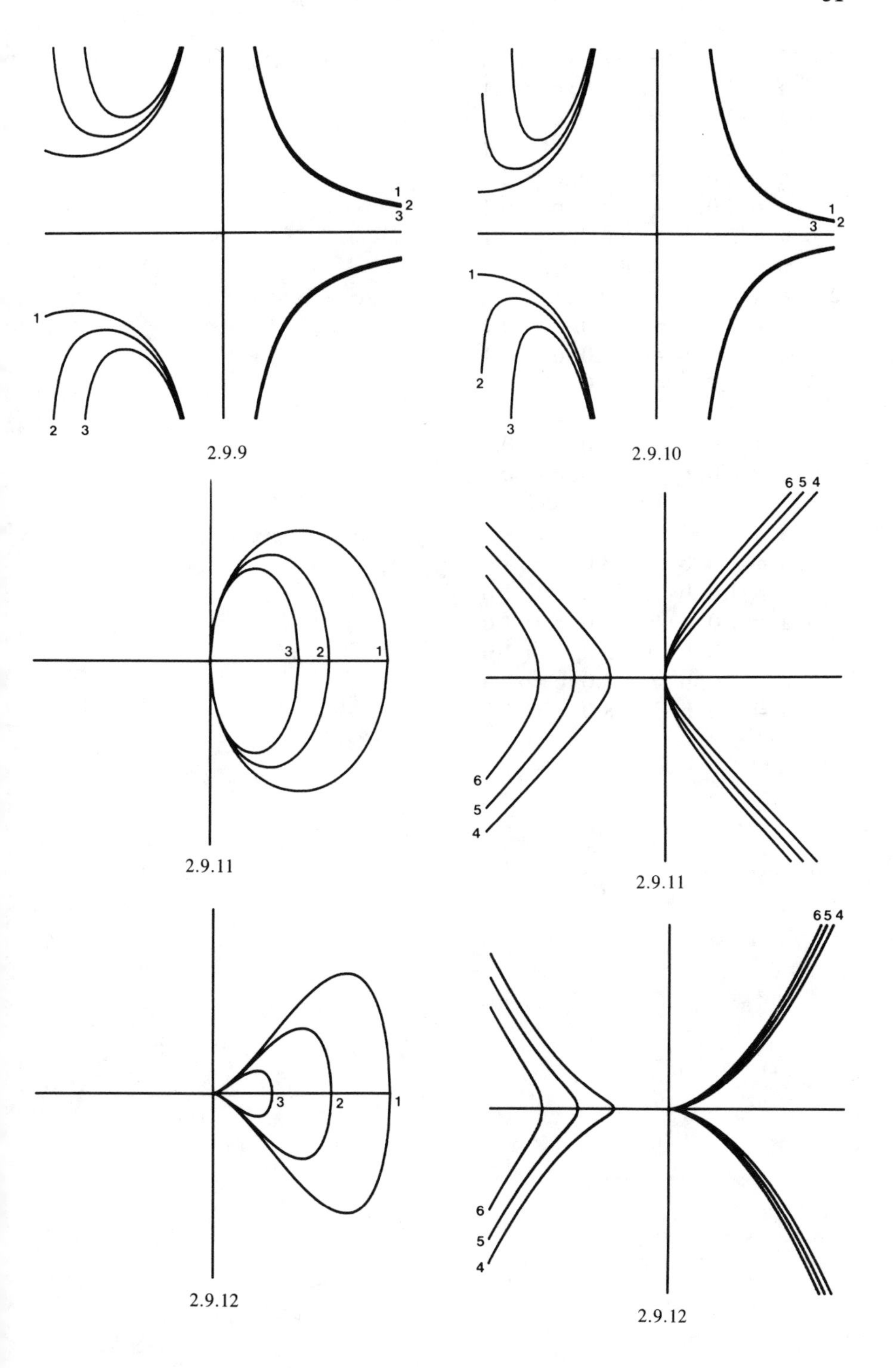

2.9.9

2.9.10

2.9.11

2.9.11

2.9.12

2.9.12

2.9.13. $y = c(a + bx)^{1/2}/x^{1/2}$ $xy^2 - c^2bx - c^2a = 0$

1. $a = 2.0$, $b = 4.0$, $c = 0.1$
2. $a = 2.0$, $b = 6.0$, $c = 0.1$
3. $a = 2.0$, $b = 8.0$, $c = 0.1$
4. $a = 2.0$, $b = -2.0$, $c = 0.1$
5. $a = 2.0$, $b = -3.0$, $c = 0.1$
6. $a = 2.0$, $b = -6.0$, $c = 0.1$

2.9.14. $y = c(a + bx)^{1/2}/x^{3/2}$ $x^3y - c^2bx - c^2a = 0$

1. $a = 2.0$, $b = 4.0$, $c = 0.1$
2. $a = 2.0$, $b = 6.0$, $c = 0.1$
3. $a = 2.0$, $b = 8.0$, $c = 0.1$
4. $a = 2.0$, $b = -2.0$, $c = 0.1$
5. $a = 2.0$, $b = -3.0$, $c = 0.1$
6. $a = 2.0$, $b = -6.0$, $c = 0.1$

2.9.15. $y = cx^{1/2}/(a + bx)^{1/2}$ $ay^2 + bxy^2 - c^2x = 0$

1. $a = 1.0$, $b = 3.0$, $c = 1.0$
2. $a = 1.0$, $b = 5.0$, $c = 1.0$
3. $a = 1.0$, $b = 8.0$, $c = 1.0$
4. $a = 4.0$, $b = -2.0$, $c = 1.0$
5. $a = 4.0$, $b = -4.0$, $c = 1.0$
6. $a = 4.0$, $b = -8.0$, $c = 1.0$

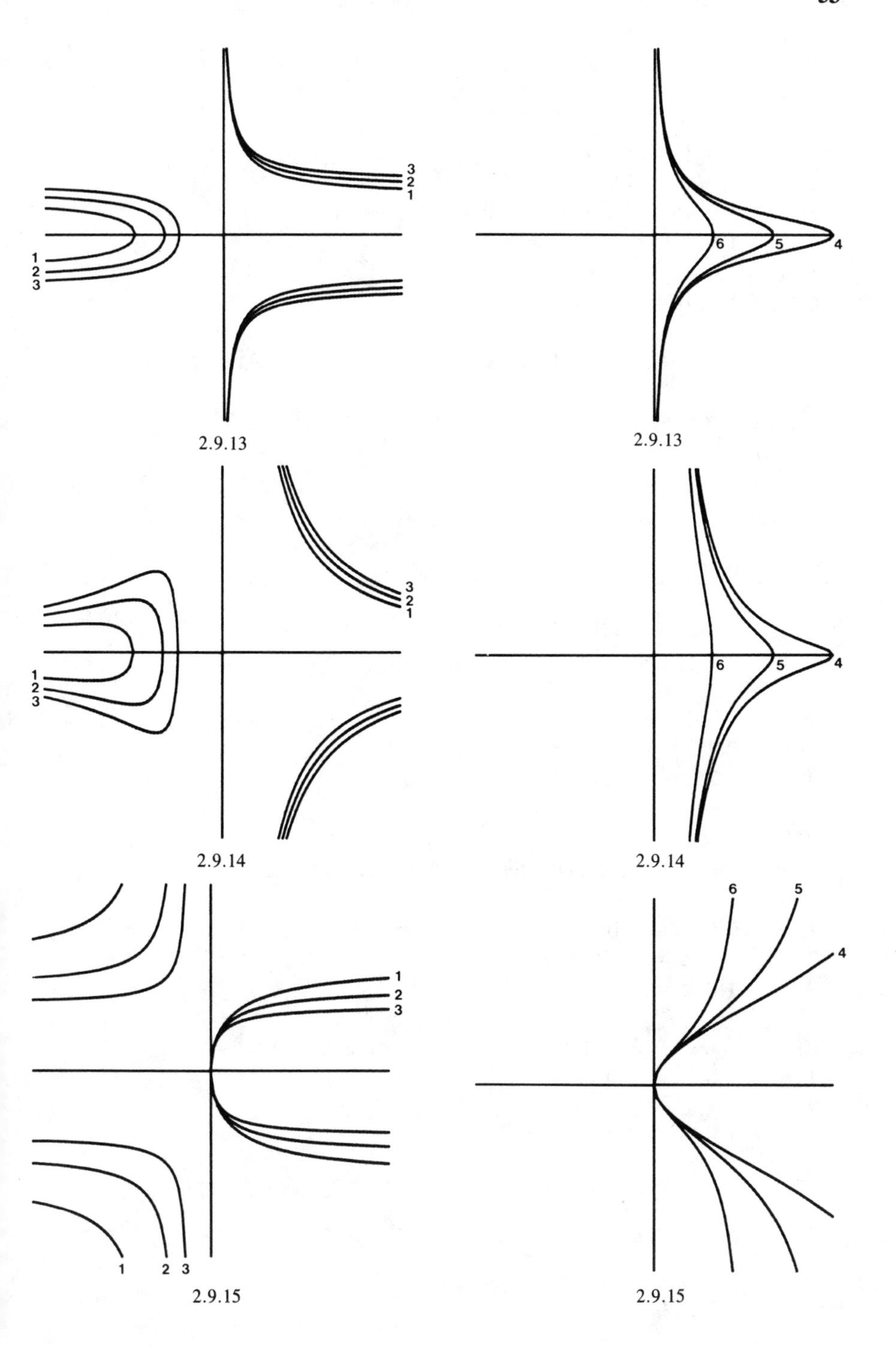

2.9.13

2.9.13

2.9.14

2.9.14

2.9.15

2.9.15

2.9.16. $y = cx^{3/2}/(a + bx)^{1/2}$ $\quad ay^2 + bxy^2 - c^2x^3 = 0$
Special case: $b = -a$ gives "cissoid of Diocles"

1. a = 1.0, b = 3.0, c = 1.0
2. a = 1.0, b = 4.0, c = 1.0
3. a = 1.0, b = 8.0, c = 1.0
4. a = 1.0, b = −0.5, c = 1.0
5. a = 1.0, b = −1.0, c = 1.0
6. a = 1.0, b = −2.0, c = 1.0

2.10. FUNCTIONS WITH $(a^2 - x^2)^{1/2}$ AND x^m

2.10.1. $y = (a^2 - x^2)^{1/2}$ $\quad y^2 + x^2 - a^2 = 0$
"Circle"

1. a = 1.00
2. a = 0.75
3. a = 0.50

2.10.2. $y = (a^2 - b^2x^2)^{1/2}$ $\quad y^2 + b^2x^2 - a^2 = 0$
"Ellipse"

1. a = 0.75, b = 0.75
2. a = 0.75, b = 1.00
3. a = 0.75, b = 1.50

2.10.3. $y = cx(a^2 - x^2)^{1/2}$ $\quad y^2 - c^2a^2x^2 + c^2x^4 = 0$
"Eight curve" (also called "lemniscate of Gerono")

1. a = 0.6, c = 1.0
2. a = 0.8, c = 1.0
3. a = 1.0, c = 1.0

2.10.4. $y = cx^2(a^2 - x^2)^{1/2}$ $\quad y^2 - c^2a^2x^4 + c^2x^6 = 0$
1. a = 0.6, c = 2.0
2. a = 0.8, c = 2.0
3. a = 1.0, c = 2.0

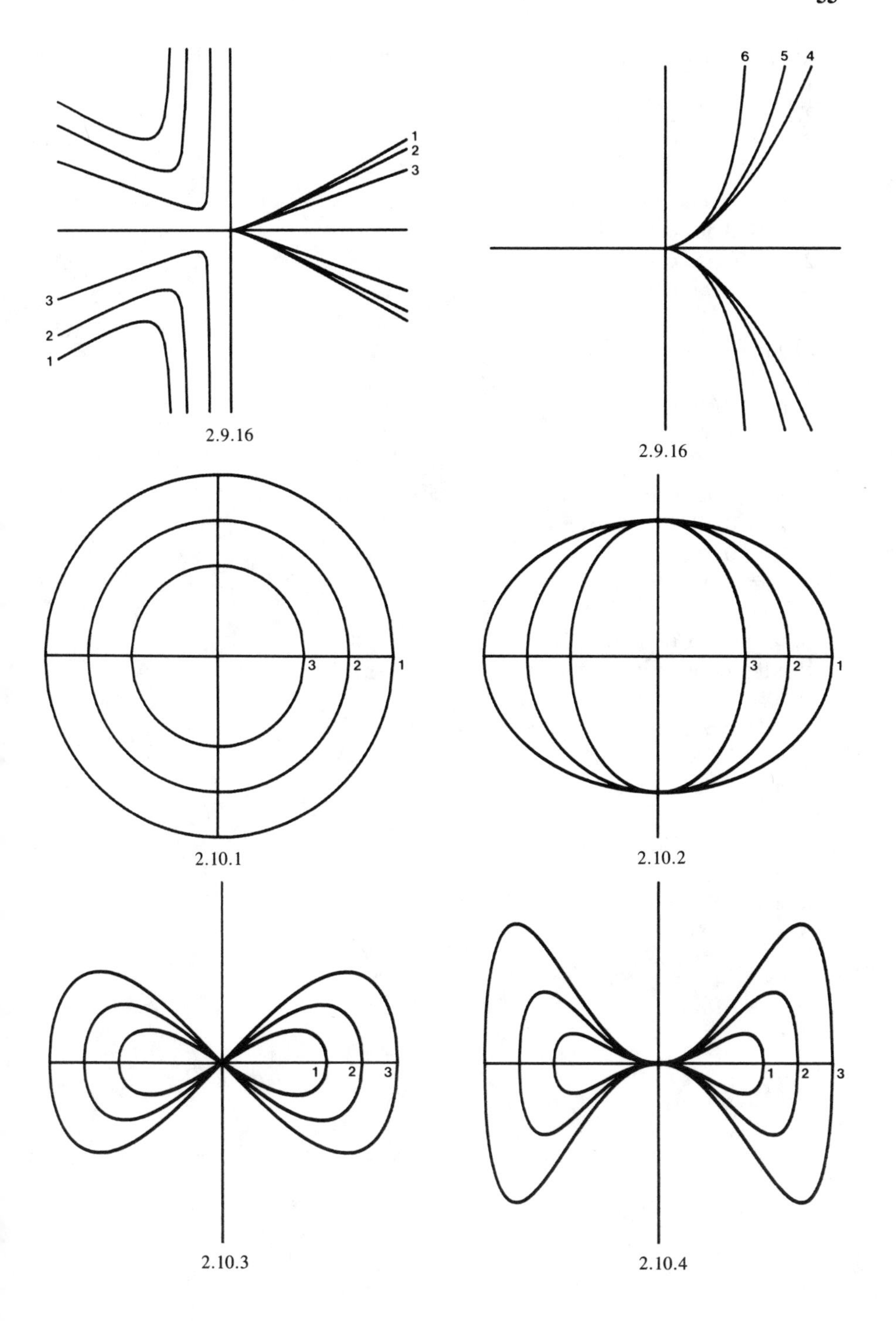

2.9.16

2.9.16

2.10.1

2.10.2

2.10.3

2.10.4

2.10.5. $y = c(a^2 - x^2)^{1/2}/x$ — $x^2y^2 + c^2x^2 - c^2a^2 = 0$

1. $a = 0.50$, $c = 0.1$
2. $a = 0.75$, $c = 0.1$
3. $a = 1.00$, $c = 0.1$

2.10.6. $y = c(a^2 - x^2)^{1/2}/x^2$ — $x^4y^2 + c^2x^2 - c^2a^2 = 0$

1. $a = 0.50$, $c = 0.1$
2. $a = 0.75$, $c = 0.1$
3. $a = 1.00$, $c = 0.1$

2.10.7. $y = c/(a^2 - x^2)^{1/2}$ — $a^2y^2 - x^2y^2 - c^2 = 0$

1. $a = 0.50$, $c = 0.1$
2. $a = 0.75$, $c = 0.1$
3. $a = 1.00$, $c = 0.1$

2.10.8. $y = c/(x(a^2 - x^2)^{1/2})$ — $a^2x^2y^2 - x^4y^2 - c^2 = 0$

1. $a = 0.50$, $c = 0.1$
2. $a = 0.75$, $c = 0.1$
3. $a = 1.00$, $c = 0.1$

2.10.9. $y = cx/(a^2 - x^2)^{1/2}$ — $a^2y^2 - x^2y^2 - c^2x^2 = 0$

''Bullet nose curve''

1. $a = 0.50$, $c = 0.4$
2. $a = 0.75$, $c = 0.4$
3. $a = 1.00$, $c = 0.4$

2.10.10. $y = cx^2/(a^2 - x^2)^{1/2}$ — $a^2y^2 - x^2y^2 - c^2x^4 = 0$

1. $a = 0.50$, $c = 0.4$
2. $a = 0.75$, $c = 0.4$
3. $a = 1.00$, $c = 0.4$

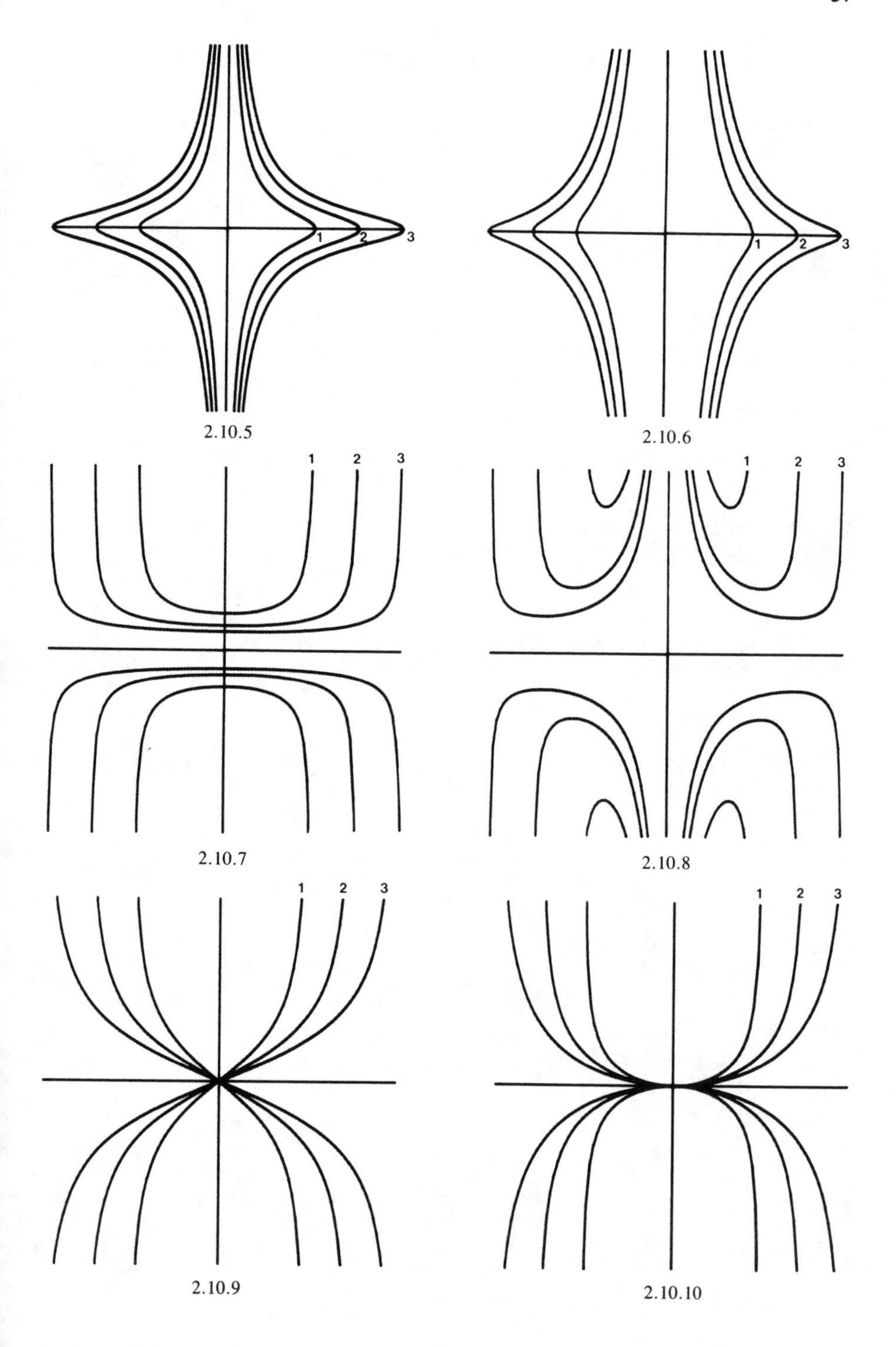

2.10.5

2.10.6

2.10.7

2.10.8

2.10.9

2.10.10

2.11. FUNCTIONS WITH $(x^2 - a^2)^{1/2}$ AND x^m

2.11.1. $y = (x^2 - a^2)^{1/2}$ $y^2 - x^2 + a^2 = 0$
"Simple hyperbola"

1. a = 0.1
2. a = 0.3
3. a = 0.5

2.11.2. $y = (b^2x^2 - a^2)^{1/2}$ $y^2 - b^2x^2 + a^2 = 0$
"Hyperbola"

1. a = 0.1, b = 0.5
2. a = 0.3, b = 1.0
3. a = 0.5, b = 1.5

2.11.3. $y = cx(x^2 - a^2)^{1/2}$ $y^2 - c^2x^4 + c^2a^2x^2 = 0$
"Kampyle of Eudoxus"

1. a = 0.1, c = 1.0
2. a = 0.4, c = 1.0
3. a = 0.7, c = 1.0

2.11.4. $y = cx^2(x^2 - a^2)^{1/2}$ $y^2 - c^2x^6 + c^2a^2x^4 = 0$
1. a = 0.1, c = 1.0
2. a = 0.4, c = 1.0
3. a = 0.7, c = 1.0

2.11.5. $y = c(x^2 - a^2)^{1/2}/x$ $x^2y^2 - c^2x^2 + c^2a^2 = 0$
1. a = 0.1, c = 1.0
2. a = 0.3, c = 1.0
3. a = 0.5, c = 1.0

2.11.6. $y = c(x^2 - a^2)^{1/2}/x^2$ $x^4y^2 - c^2x^2 + c^2a^2 = 0$
1. a = 0.1, c = 0.2
2. a = 0.2, c = 0.2
3. a = 0.3, c = 0.2

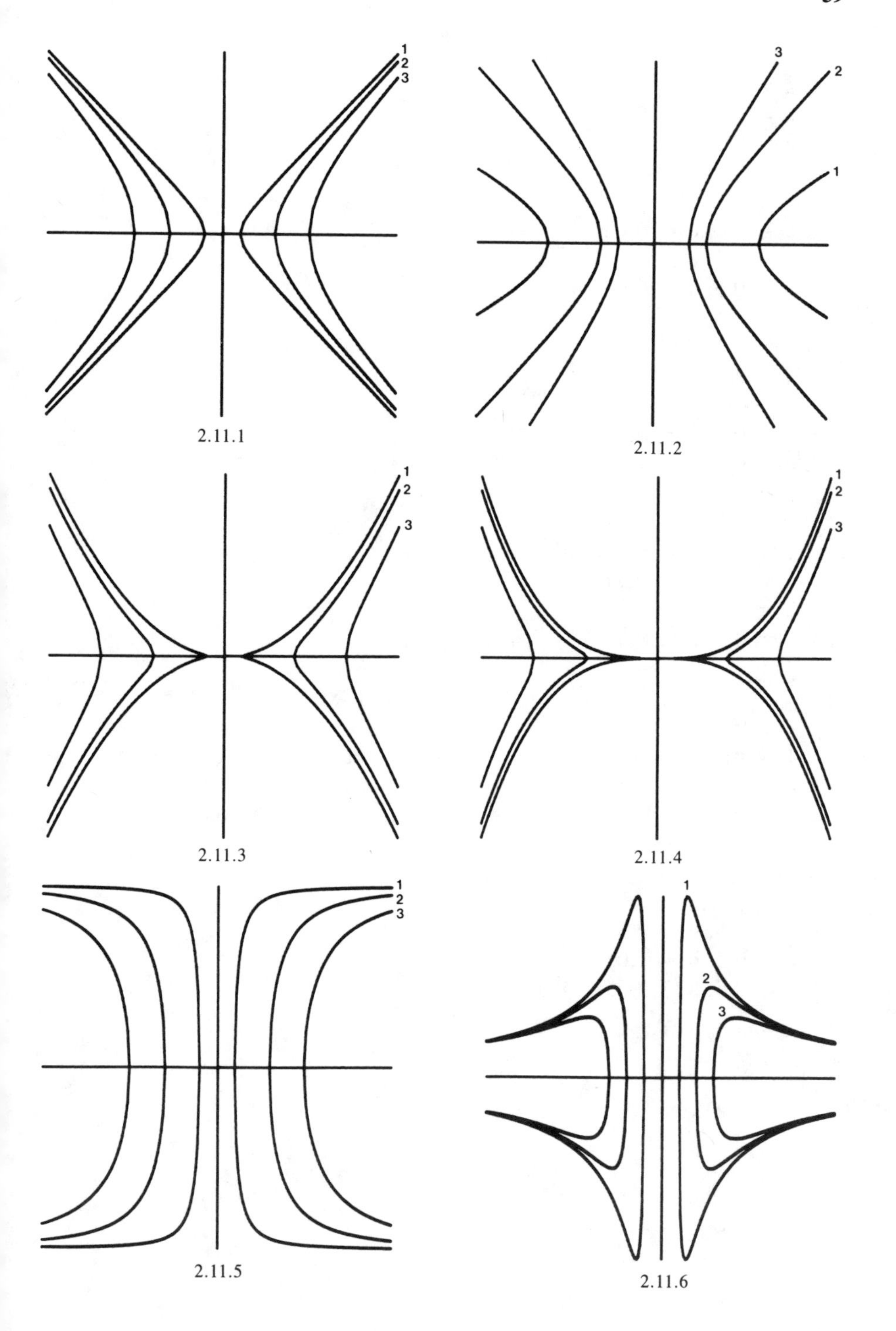

2.11.1

2.11.2

2.11.3

2.11.4

2.11.5

2.11.6

2.11.7. $y = c/(x^2 - a^2)^{1/2}$ $x^2y^2 - a^2y^2 - c^2 = 0$

1. $a = 0.1, c = 0.1$
2. $a = 0.3, c = 0.1$
3. $a = 0.5, c = 0.1$

2.11.8. $y = c/(x(x^2 - a^2)^{1/2})$ $x^4y^2 - a^2x^2y^2 - c^2 = 0$

1. $a = 0.1, c = 0.02$
2. $a = 0.3, c = 0.02$
3. $a = 0.5, c = 0.02$

2.11.9. $y = cx/(x^2 - a^2)^{1/2}$ $x^2y^2 - a^2y^2 - c^2x^2 = 0$

"Cross curve"

1. $a = 0.1, c = 0.2$
2. $a = 0.3, c = 0.2$
3. $a = 0.5, c = 0.2$

2.11.10. $y = cx^2/(x^2 - a^2)^{1/2}$ $x^2y^2 - a^2y^2 - c^2x^4 = 0$

1. $a = 0.2, c = 0.5$
2. $a = 0.3, c = 0.5$
3. $a = 0.4, c = 0.5$

2.12. FUNCTIONS WITH $(a^2 + x^2)^{1/2}$ AND x^m

2.12.1. $y = c(a^2 + x^2)^{1/2}$ $y^2 - c^2a^2 - c^2x^2 = 0$

1. $a = 0.1, c = 0.75$
2. $a = 0.3, c = 0.75$
3. $a = 0.5, c = 0.75$

2.12.2. $y = c(a^2 + b^2x^2)^{1/2}$ $y^2 - c^2a^2 - c^2b^2x^2 = 0$

1. $a = 0.5, b = 0.5, c = 0.75$
2. $a = 0.5, b = 1.0, c = 0.75$
3. $a = 0.5, b = 2.0, c = 0.75$

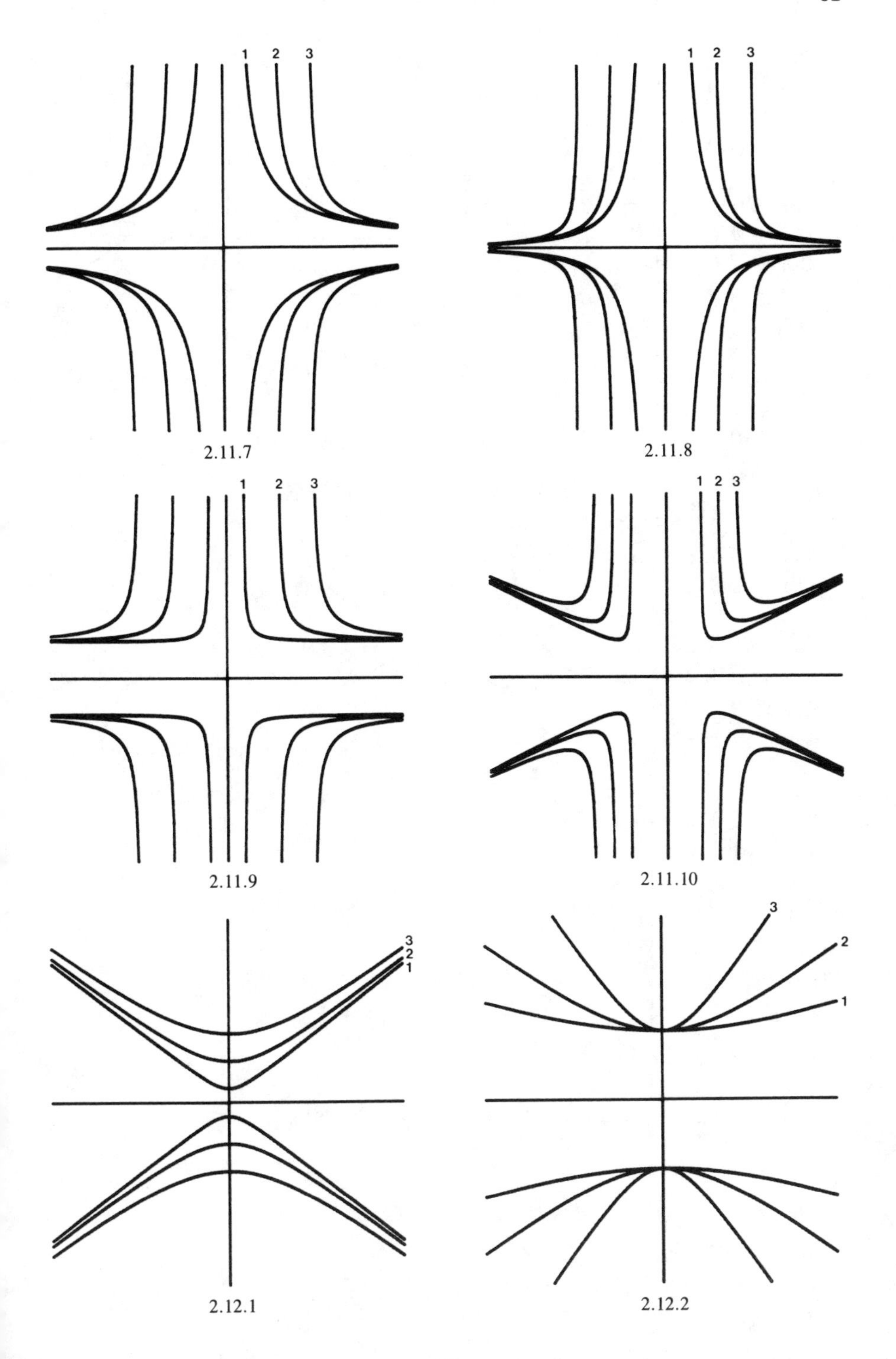

2.11.7

2.11.8

2.11.9

2.11.10

2.12.1

2.12.2

2.12.3. $y = cx(a^2 + x^2)^{1/2}$ $\qquad y^2 - c^2a^2x^2 - c^2x^4 = 0$

1. $a = 1.0,\ c = 0.5$
2. $a = 2.0,\ c = 0.5$
3. $a = 4.0,\ c = 0.5$

2.12.4. $y = cx^2(a^2 + x^2)^{1/2}$ $\qquad y^2 - c^2a^2x^4 - c^2x^6 = 0$

1. $a = 1.0,\ c = 0.5$
2. $a = 2.0,\ c = 0.5$
3. $a = 4.0,\ c = 0.5$

2.12.5. $y = c(a^2 + x^2)^{1/2}/x$ $\qquad x^2y^2 - c^2x^2 - c^2a^2 = 0$

1. $a = 0.2,\ c = 0.2$
2. $a = 0.5,\ c = 0.2$
3. $a = 0.8,\ c = 0.2$

2.12.6. $y = c(a^2 + x^2)^{1/2}/x^2$ $\qquad x^4y^2 - c^2x^2 - c^2a^2 = 0$

1. $a = 0.2,\ c = 0.2$
2. $a = 0.5,\ c = 0.2$
3. $a = 0.8,\ c = 0.2$

2.12.7. $y = c/(a^2 + x^2)^{1/2}$ $\qquad a^2y^2 + x^2y^2 - c^2 = 0$

1. $a = 0.2,\ c = 0.2$
2. $a = 0.5,\ c = 0.2$
3. $a = 0.8,\ c = 0.2$

2.12.8. $y = c/(x(a^2 + x^2)^{1/2})$ $\qquad a^2x^2y^2 + x^4y^2 - c^2 = 0$

1. $a = 1.0,\ c = 0.5$
2. $a = 2.0,\ c = 0.5$
3. $a = 4.0,\ c = 0.5$

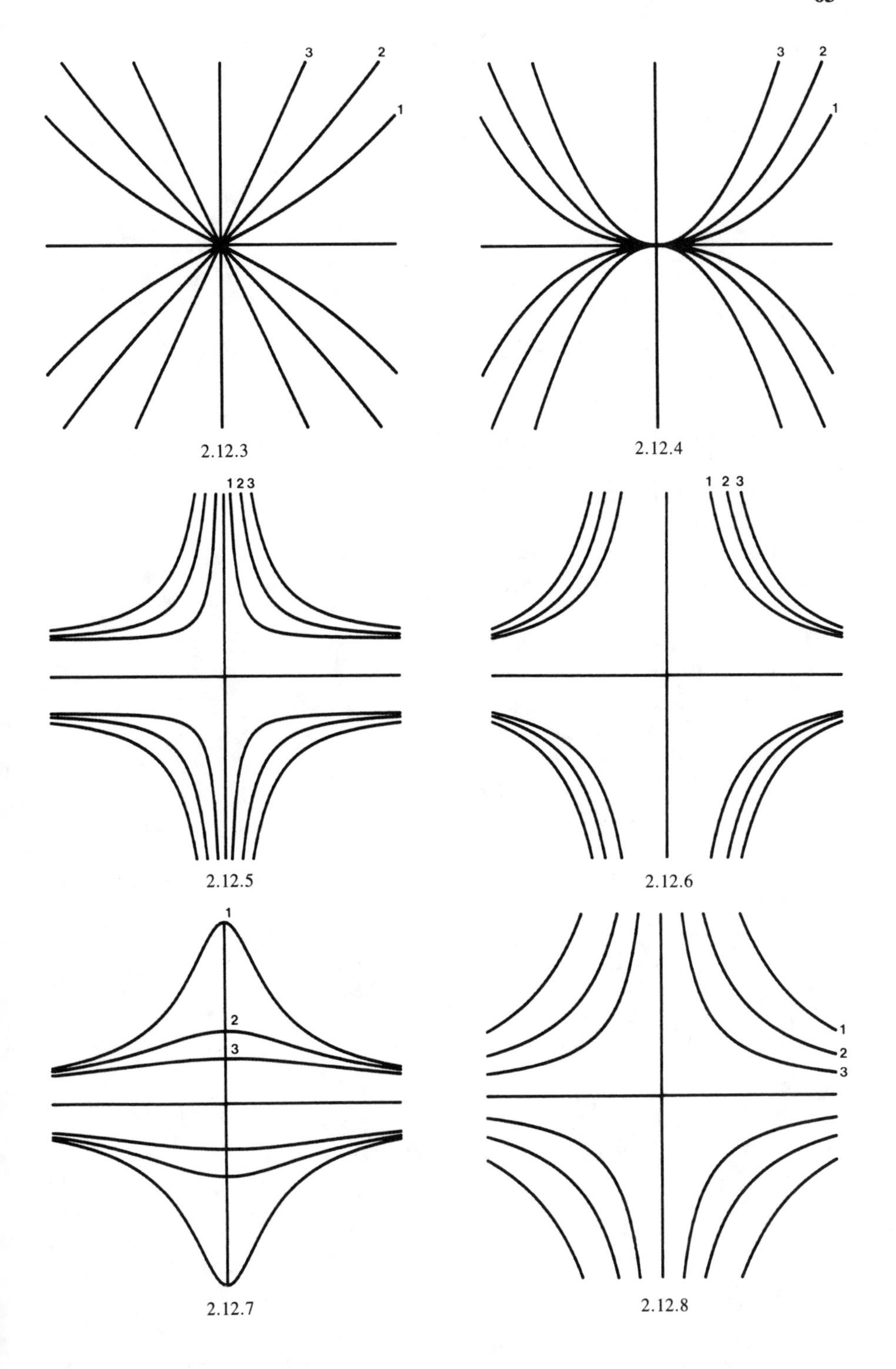

2.12.3

2.12.4

2.12.5

2.12.6

2.12.7

2.12.8

2.12.9. $y = cx/(a^2 + x^2)^{1/2}$ $a^2y^2 + x^2y^2 - c^2x^2 = 0$

1. $a = 0.5, c = 1.0$
2. $a = 1.0, c = 1.0$
3. $a = 2.0, c = 1.0$

2.12.10. $y = cx^2/(a^2 + x^2)^{1/2}$ $a^2y^2 + x^2y^2 - c^2x^4 = 0$

1. $a = 0.1, c = 1.0$
2. $a = 0.5, c = 1.0$
3. $a = 1.0, c = 1.0$

2.13. MISCELLANEOUS ALGEBRAIC FUNCTIONS

2.13.1. $y = c(a + x)/(b - x)$ $by - xy - cx - ca = 0$

1. $a = 0.5, b = 0.3, c = 0.1$
2. $a = 0.5, b = 0.6, c = 0.1$
3. $a = 0.1, b = 0.6, c = 0.1$
4. $a = -0.5, b = 0.3, c = 0.1$
5. $a = -0.5, b = 0.6, c = 0.1$
6. $a = -0.1, b = 0.6, c = 0.1$

2.13.2. $y = c(a + x)^{1/2}/(b - x)^{1/2}$ $by^2 - xy^2 - c^2x - c^2a = 0$

1. $a = 1.0, b = 0.8, c = 0.2$
2. $a = 1.0, b = 0.3, c = 0.2$
3. $a = 0.5, b = 0.3, c = 0.2$
4. $a = -0.1, b = 0.5, c = 0.2$
5. $a = -0.1, b = 0.8, c = 0.2$
6. $a = -0.3, b = 0.8, c = 0.2$

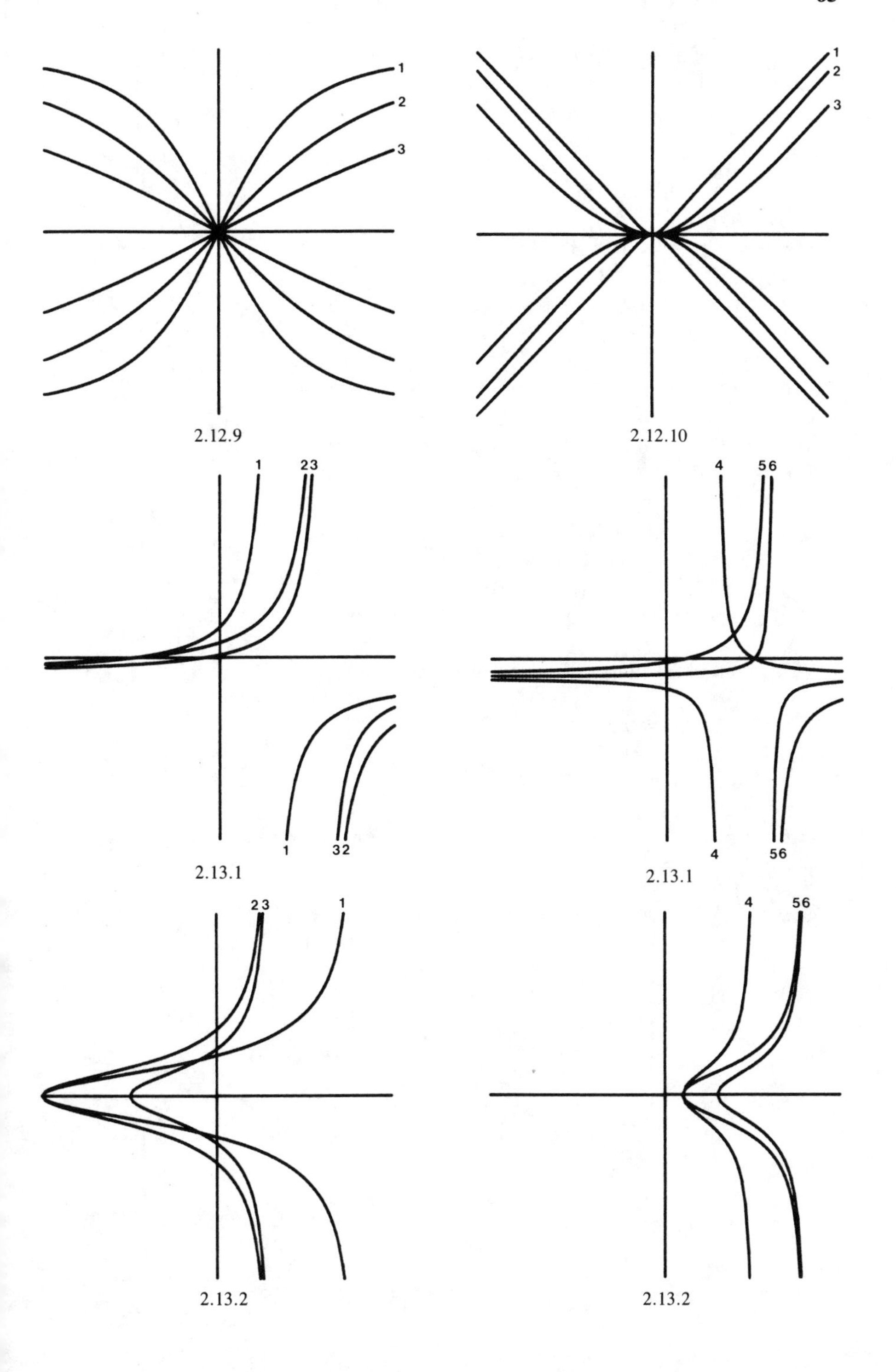

2.12.9

2.12.10

2.13.1

2.13.1

2.13.2

2.13.2

2.13.3. $y = cx(a + x)^{1/2}/(b - x)^{1/2}$ $\quad by^2 - xy^2 - c^2ax^2 - c^2x^3 = 0$

Special cases

a = b gives "right strophoid"

a = 3b gives "trisectrix of Maclaurin"

1. a = 1.0, b = 0.3, c = 0.4
2. a = 1.0, b = 0.8, c = 0.4
3. a = −0.3, b = 0.8, c = 0.4
4. a = 1.0, b = −0.1, c = 0.4
5. a = 1.0, b = −0.3, c = 0.4
6. a = 0.6, b = −0.3, c = 0.4

2.13.4. $y = (c/x)(a + x)^{1/2}/(b - x)^{1/2}$ $\quad bx^2y^2 - x^3y^2 - c^2x - c^2a = 0$

1. a = 1.0, b = 0.3, c = 0.05
2. a = 1.0, b = 0.8, c = 0.05
3. a = −0.3, b = 0.8, c = 0.05
4. a = 1.0, b = −0.1, c = 0.05
5. a = 1.0, b = −0.3, c = 0.05
6. a = 0.6, b = −0.3, c = 0.05

2.13.5. $y = [(a^2x^2/(x - b)^2) - x^2]^{1/2}$ $\quad b^2y^2 - 2bxy^2 + x^2y^2 + b^2x^2 - a^2x^2 - 2bx^3 + x^4 = 0$

"Conchoid of Nicomedes" (also called "cochloid")

1. a = 0.50, b = 0.50
2. a = 0.50, b = 0.25
3. a = 0.25, b = 0.50

2.13.6. $y = c(a^2 + x^2)/(b^2 - x^2)$ $\quad b^2y - x^2y - cx^2 - ca^2 = 0$

1. a = 1.0, b = 0.7, c = 0.1
2. a = 1.0, b = 0.5, c = 0.1
3. a = 0.7, b = 0.5, c = 0.1

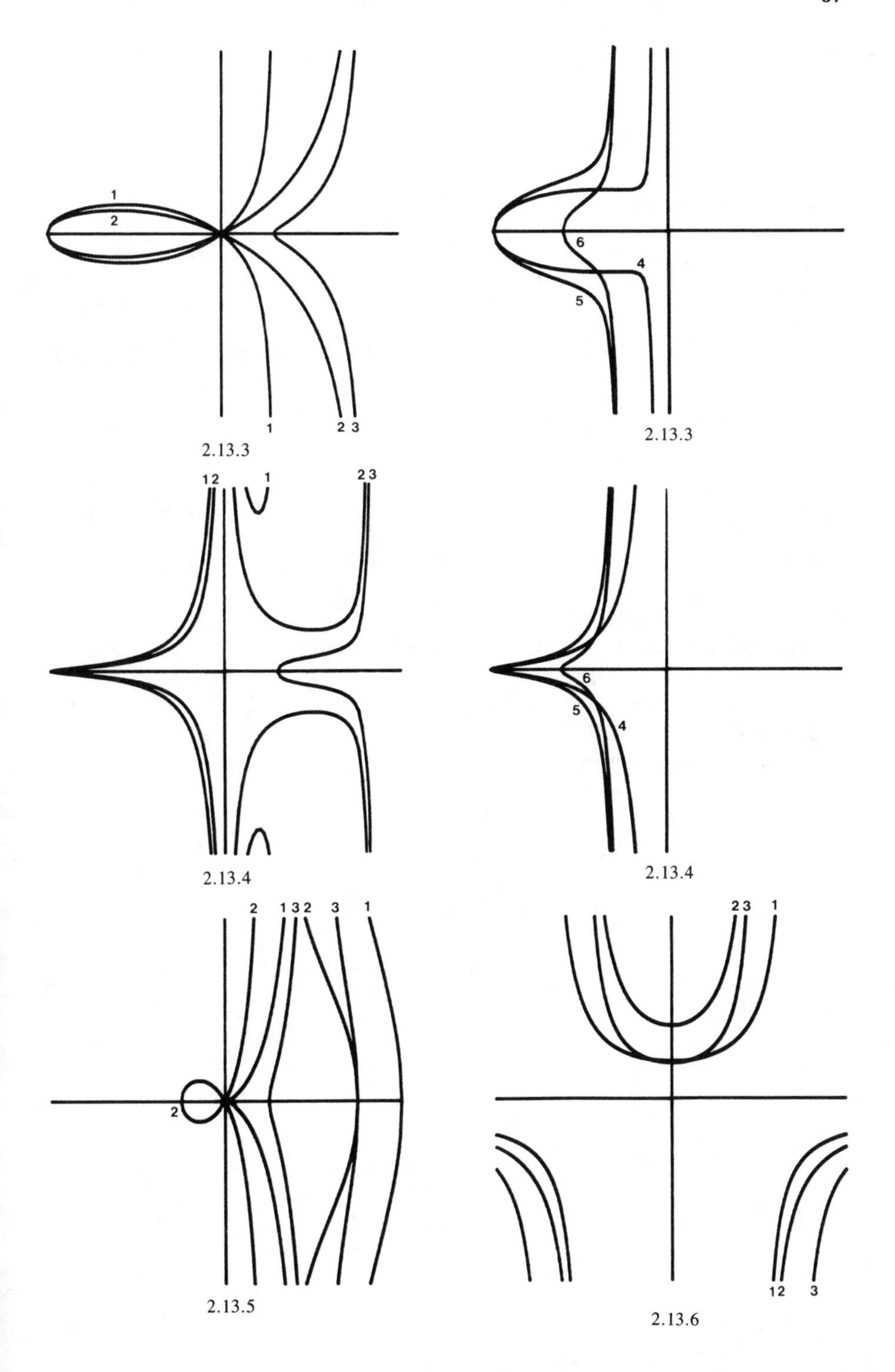

2.13.3

2.13.3

2.13.4

2.13.4

2.13.5

2.13.6

2.13.7. $y = x - (x^2 - a^2)^{1/2}$ $y^2 - 2xy + a^2 = 0$

1. $a = 0.3$
2. $a = 0.5$
3. $a = 0.7$

2.13.8. $y = x - (a^2 + x^2)^{1/2}$ $y^2 - 2xy - a^2 = 0$

1. $a = 0.3$
2. $a = 0.5$
3. $a = 0.7$

2.13.9. $y = \dfrac{(a^2 - x^2)[2a \pm (a^2 - x^2)^{1/2}]}{(3a^2 + x^2)}$ $y^2(a^2 - x^2) - (x^2 + 2ay - a^2)^2 = 0$

"Bicorn"

1. $a = 0.50$
2. $a = 0.75$
3. $a = 1.00$

2.13.10. $y = (1 - |x/a|^{n/m})^{m/n}$

"Hyperellipse" for $n/m > 2$, "hypoellipse" for $n/m < 2$

1. $a = 0.5$, $n/m = 1/3$
2. $a = 0.5$, $n/m = 2/3$
3. $a = 0.5$, $n/m = 1$
4. $a = 0.5$, $n/m = 3/2$
5. $a = 0.5$, $n/m = 2$
6. $a = 0.5$, $n/m = 4$

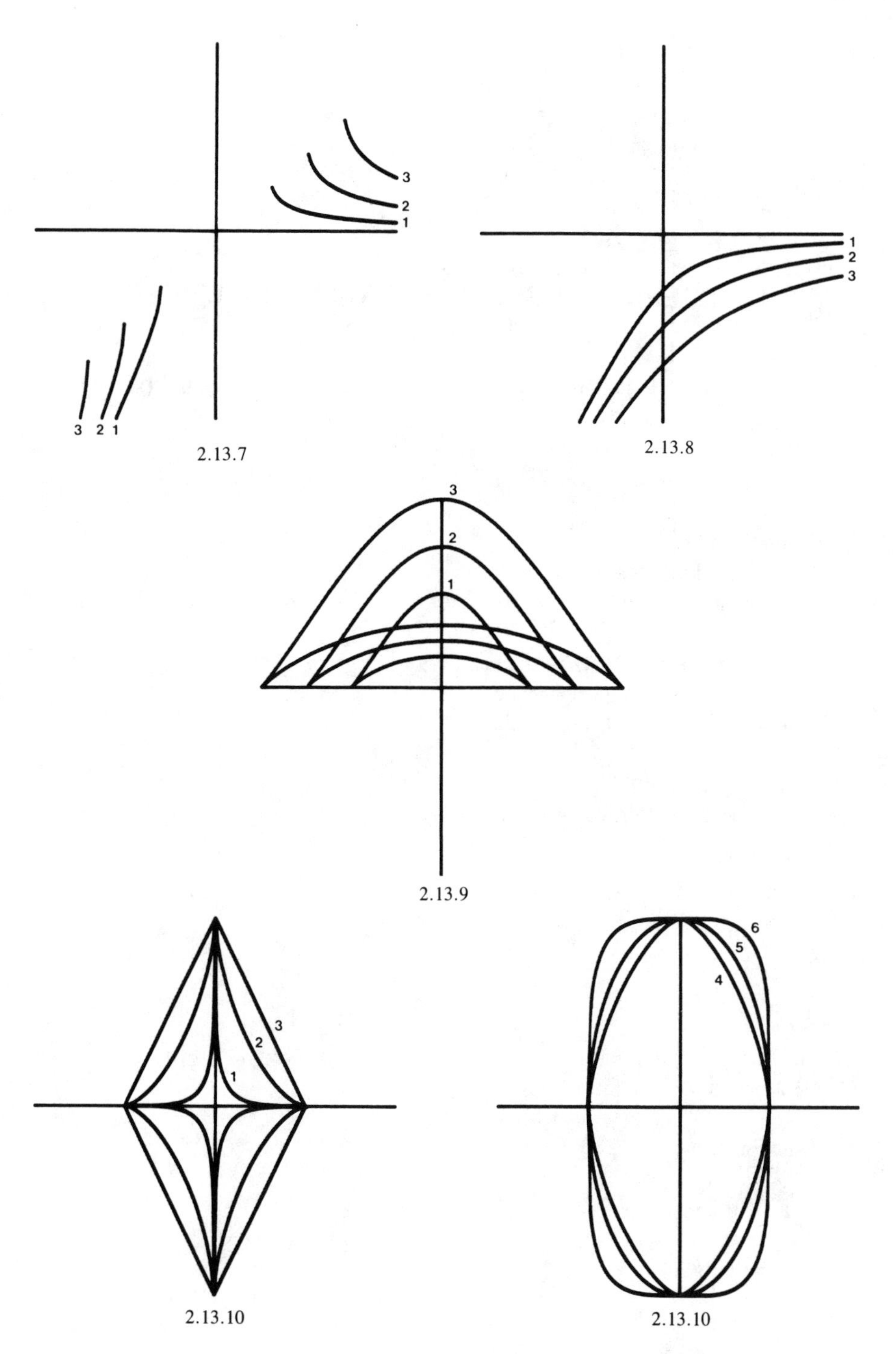

2.13.7

2.13.8

2.13.9

2.13.10

2.13.10

2.13.11. $y = (1 + |x/a|^{n/m})^{m/n}$
1. a = 0.2, n/m = 1/3, c = 0.2
2. a = 0.2, n/m = 2/3, c = 0.2
3. a = 0.2, n/m = 1, c = 0.2
4. a = 0.2, n/m = 3/2, c = 0.2
5. a = 0.2, n/m = 2, c = 0.2
6. a = 0.2, n/m = 4, c = 0.2

2.14. ALGEBRAIC FUNCTIONS EXPRESSIBLE IN POLAR COORDINATES

2.14.1. $r = c(2a\cdot\cos\theta + b)$ $\quad (x^2 + y^2 - 2acx)^2 - b^2c^2(x^2 + y^2) = 0$
"Limacon of Pascal"

Domain: $[0 < \theta < 2\pi]$

Special cases
b = 2a gives "cardioid"
b = a gives "trisectrix"

1. a = 1.0, b = 1.0, c = 0.25
2. a = 1.0, b = 2.0, c = 0.25
3. a = 0.5, b = 2.0, c = 0.25

2.14.2. $r^2 = a^2\cos2\theta$ $\quad (x^2 + y^2)^2 - a^2|x^2 - y^2| = 0$
"Lemniscate of Bernoulli"

Domain: $[0 < \theta < 2\pi]$

1. a = 0.50
2. a = 0.75
3. a = 1.00

2.14.3. $r = a\cdot\cot\theta$ $\quad (x^2 + y^2)y^2 - a^2x^2 = 0$
"Kappa curve"

Domain: $[0 < \theta < 2\pi]$

1. a = 0.2
2. a = 0.6
3. a = 1.0

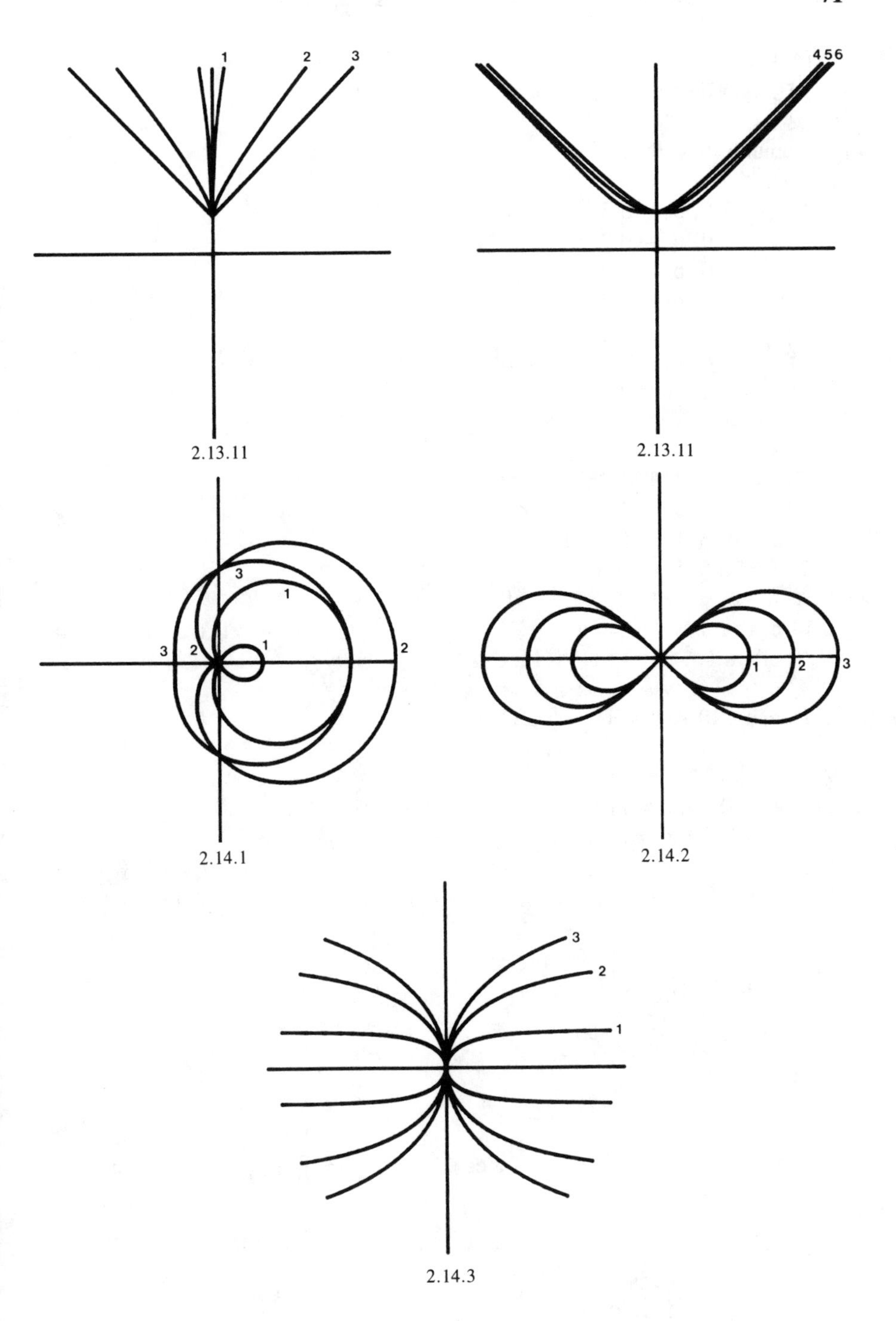

2.13.11

2.13.11

2.14.1

2.14.2

2.14.3

2.14.4. $r^2 = cb(a - b\cdot\sin^2\theta)$ — $(x^2 + y^2)^2 - cb[ax^2 + (a - b)y^2] = 0$

"Hippopede curve"

Domain: $[0 < \theta < 2\pi]$

1. a = 1.0, b = 0.5, c = 1.0
2. a = 1.0, b = 0.8, c = 1.0
3. a = 1.0, b = 1.0, c = 1.0
4. a = 0.5, b = 1.0, c = 1.0

2.14.5. $r^2 = (a^2\sin^2\theta - b^2\cos^2\theta)/(\sin^2\theta - \cos^2\theta)$ — $y^2(y^2 - a^2) - x^2(x^2 - b^2) = 0$

"Devil's curve"

Domain: $[0 < \theta < 2\pi]$

1. a = 0.3, b = 0.5
2. a = 0.5, b = 0.6

2.14.6. $r = \cos\theta(4a\cdot\sin^2\theta - b)$ — $y^2[1 + (b - 4a)x] + x^2(1 + b) = 0$

"Folium"

Domain: $[0 < \theta < \pi]$

1. a = 0.25, b = 1.0
2. a = 0.50, b = 1.0
3. a = 1.00, b = 1.0

2.14.7. $r = a\cdot\sin\theta\cdot\cos^2\theta$ — $(x^2 + y^2)^2 - ax^2y = 0$

"Bifolia"

Domain: $[0 < \theta < \pi]$

1. a = 1.0
2. a = 2.0
3. a = 3.0

2.14.8. $r^2 = (b^4 - a^4\sin^2 2\theta)^{1/2} + a^2\cos 2\theta$ — $(x^2 + y^2 + a^2)^2 - 4a^2x^2 - b^2 = 0$

"Cassinian oval"

Domain: $[0 < \theta < 2\pi]$

1. a = 0.45, b = 0.5
2. a = 0.50, b = 0.5
3. a = 0.55, b = 0.5

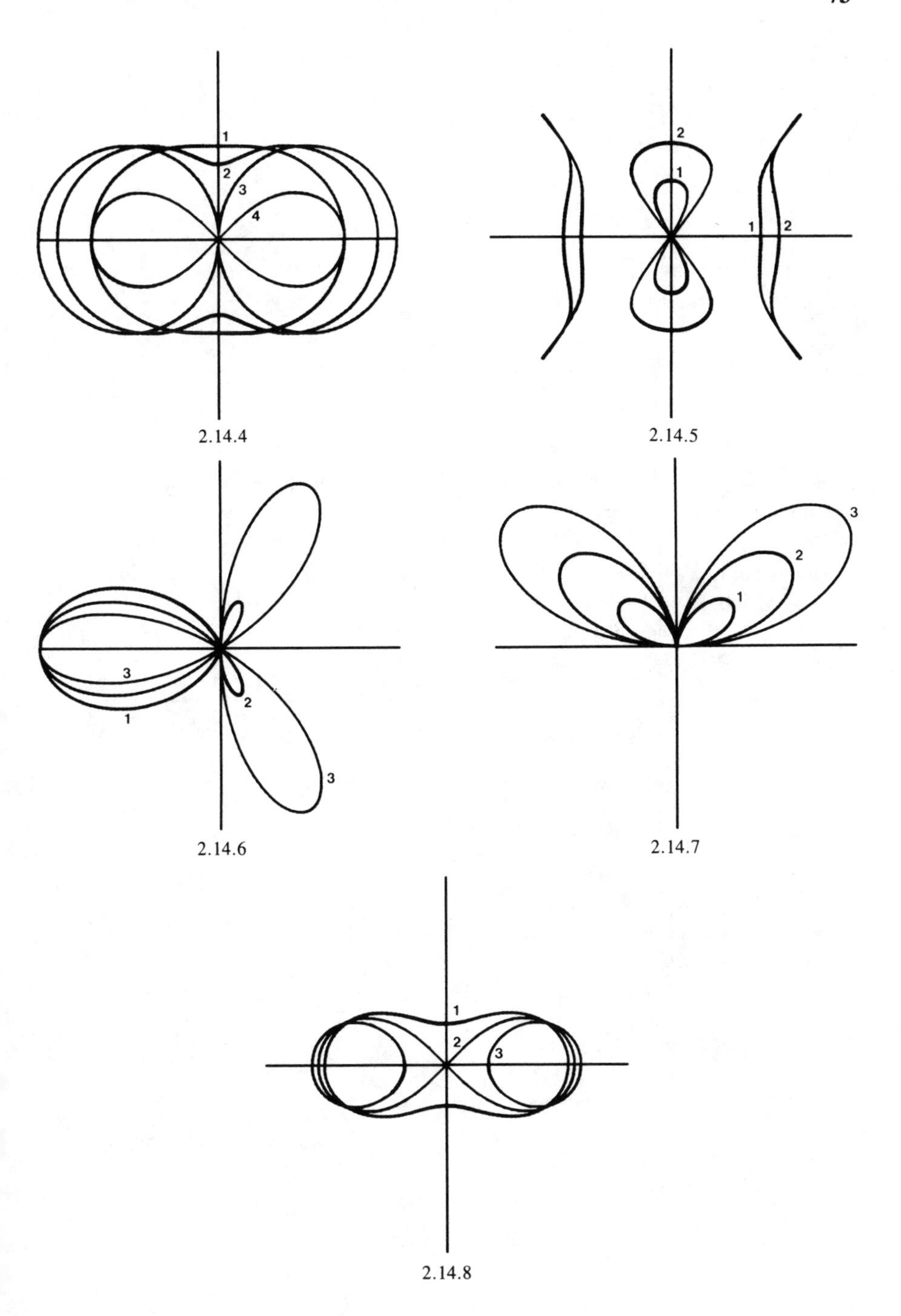

2.14.4

2.14.5

2.14.6

2.14.7

2.14.8

2.14.9. $r = a(1 + 2\cdot\sin(\theta/2))$
"Nephroid of Freeth"

$$(x^2 + y^2)(x^2 + y^2 + a^2 - 2a)^2 - [2a(x^2 + y^2) - 2ax] = 0$$

Domain: $[0 < \theta < 4\pi]$

1. $a = 0.3$

2.14.10. $r = a\cdot\cos^3(\theta/3)$
"Cayley's sextet"

$$4(x^2 + y^2 - ax)^3 - 27\,a^2(x^2 + y^2)^2 = 0$$

Domain: $[0 < \theta < 3\pi]$

1. $a = 1.0$

2.14.11. $r = c(1 - a\cdot\cos\theta)/(1 + a\cdot\cos\theta)$

$$(x^2 + y^2 + acx)^2 - (x^2 + y^2)(c - ax)^2 = 0$$

Domain: $[0 < \theta < 2\pi]$

1. $a = 0.5$, $c = 1.00$
2. $a = 1.0$, $c = 1.00$
3. $a = 10.0$, $c = 1.00$

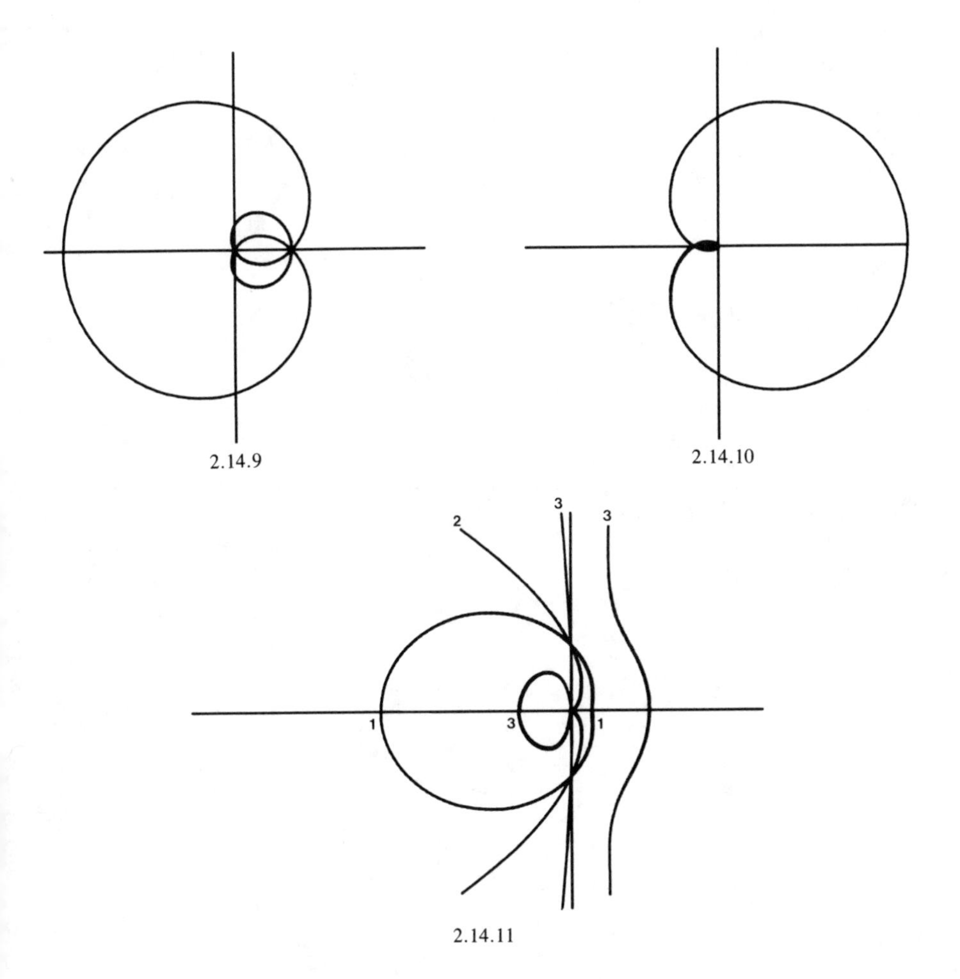

2.14.9

2.14.10

2.14.11

Chapter 3

TRANSCENDENTAL FUNCTIONS

This chapter treats the transcendental functions: trigonometric, logarithmic, and exponential. The equations found in this chapter can mostly be found in tables of integrals. Traditional or accepted names for certain curves are included wherever appropriate. A final section of the chapter comprises curves which are more easily expressed in the polar form $r = f(\theta)$ than in the Cartesian form $y = f(x)$.

3.1. TRIGONOMETRIC FUNCTIONS WITH $\sin^n(ax)$ AND $\cos^m(bx)$ (n,m INTEGERS)

3.1.1. $y = \sin(2\pi x)$

3.1.2. $y = \cos(2\pi x)$

3.1.3. $y = \tan(2\pi x)$

3.1.4. $y = \cot(2\pi x)$

3.1.5. $y = 0.25 \csc(2\pi x)$

3.1.6. $y = 0.25 \sec(2\pi x)$

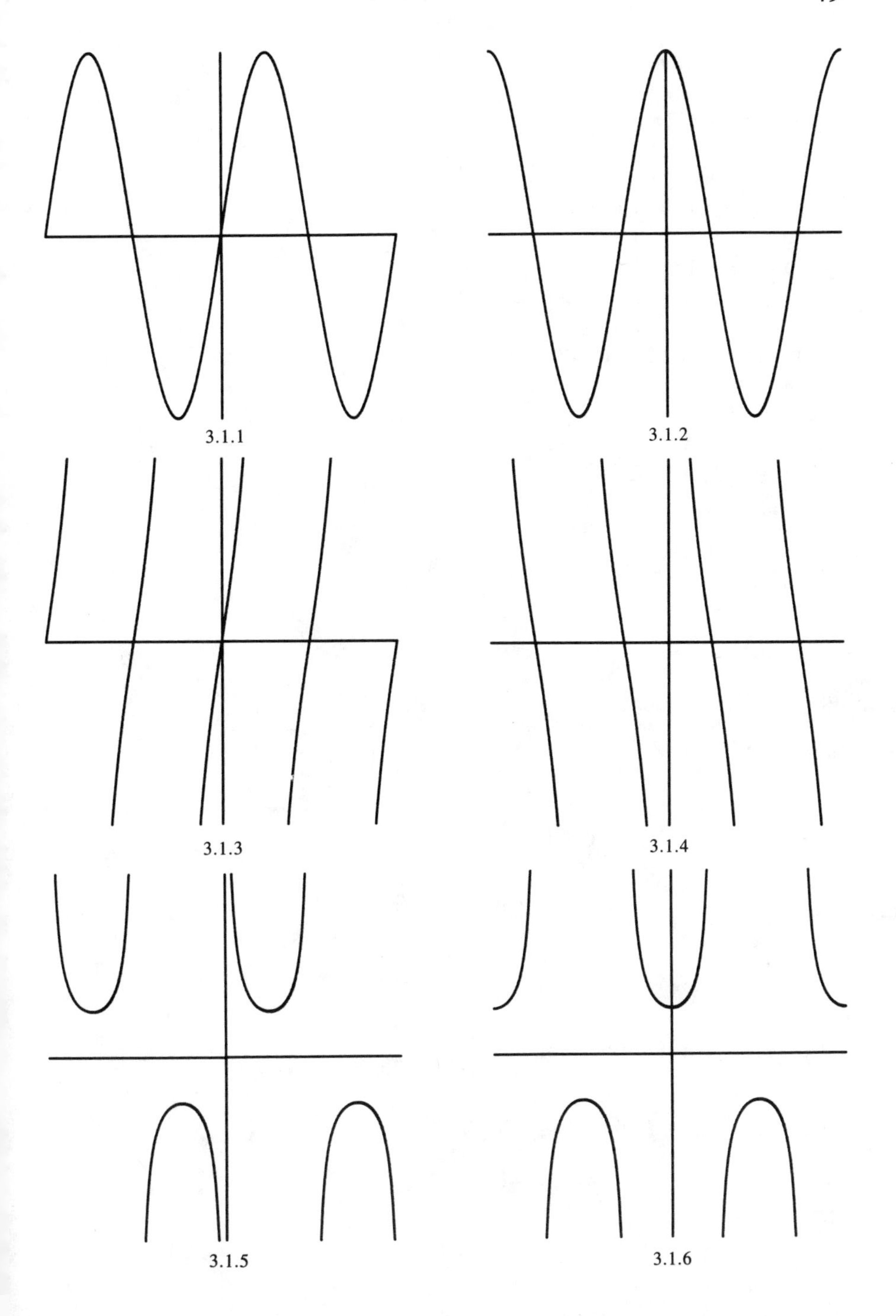
3.1.1 3.1.2 3.1.3 3.1.4 3.1.5 3.1.6

3.1.7. $y = \sin^2(2\pi x)$

3.1.8. $y = \cos^2(2\pi x)$

3.1.9. $y = \sin(2\pi ax) \cdot \sin(2\pi bx)$

1. $a = 0.5$, $b = 1.0$
2. $a = 0.5$, $b = 1.5$
3. $a = 0.5$, $b = 2.0$
4. $a = 0.5$, $b = 2.5$

3.1.10. $y = \cos(2\pi ax) \cdot \cos(2\pi bx)$

1. $a = 0.5$, $b = 1.0$
2. $a = 0.5$, $b = 1.5$
3. $a = 0.5$, $b = 2.0$
4. $a = 0.5$, $b = 2.5$

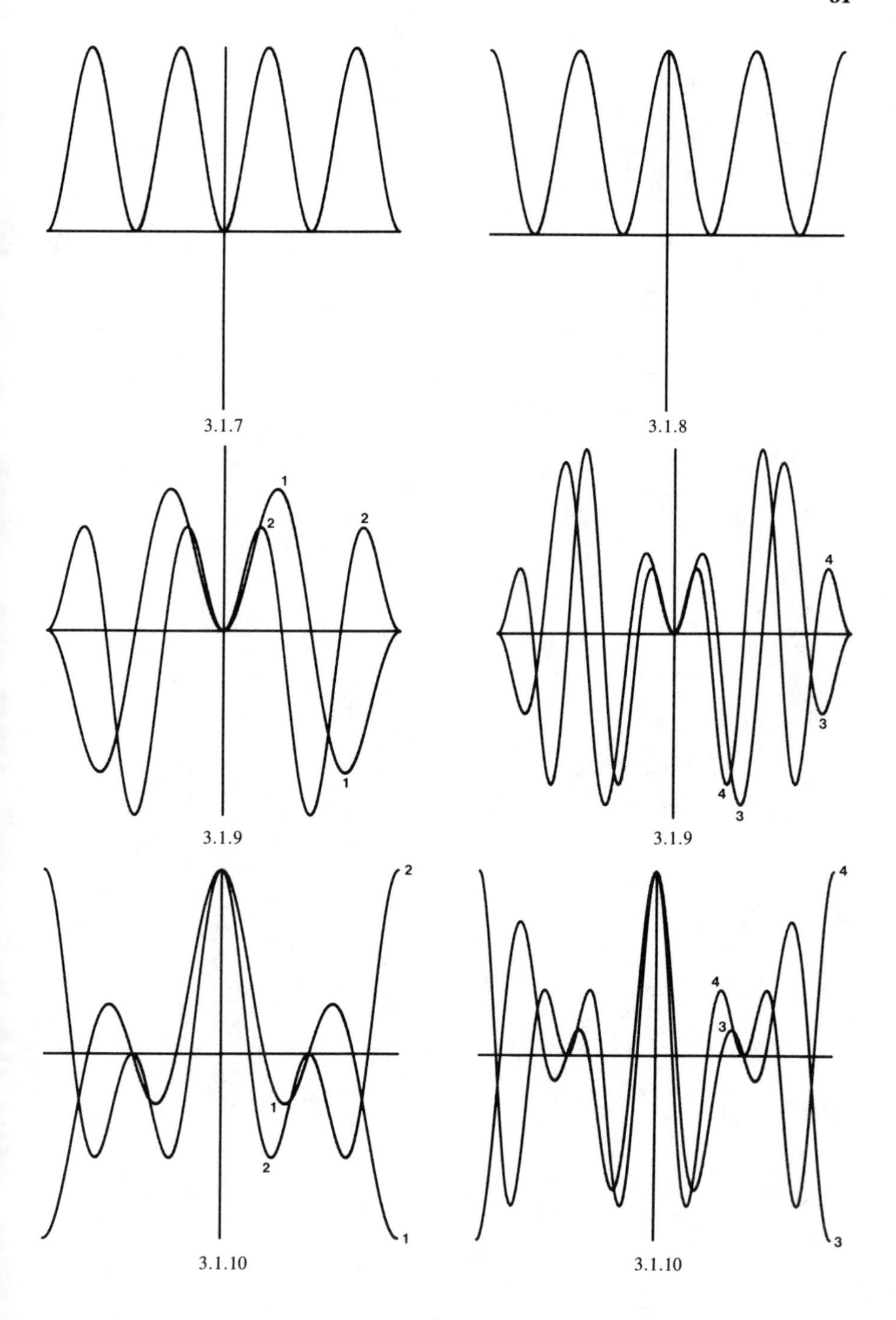

3.1.7

3.1.8

3.1.9

3.1.9

3.1.10

3.1.10

3.1.11. $y = \sin(2\pi ax)\cdot\cos(2\pi bx)$

1. $a = 0.5$, $b = 1.0$
2. $a = 0.5$, $b = 1.5$
3. $a = 0.5$, $b = 2.0$
4. $a = 0.5$, $b = 2.5$

3.1.12. $y = 2.0 \sin(2\pi x)\cdot\cos^2(2\pi x)$

3.1.13. $y = 2.0 \cos(2\pi x)\cdot\sin^2(2\pi x)$

3.1.14. $y = 0.25 \sin(2\pi x)/\cos^2(2\pi x)$

3.1.15. $y = 0.25 \sin^2(2\pi x)/\cos(2\pi x)$

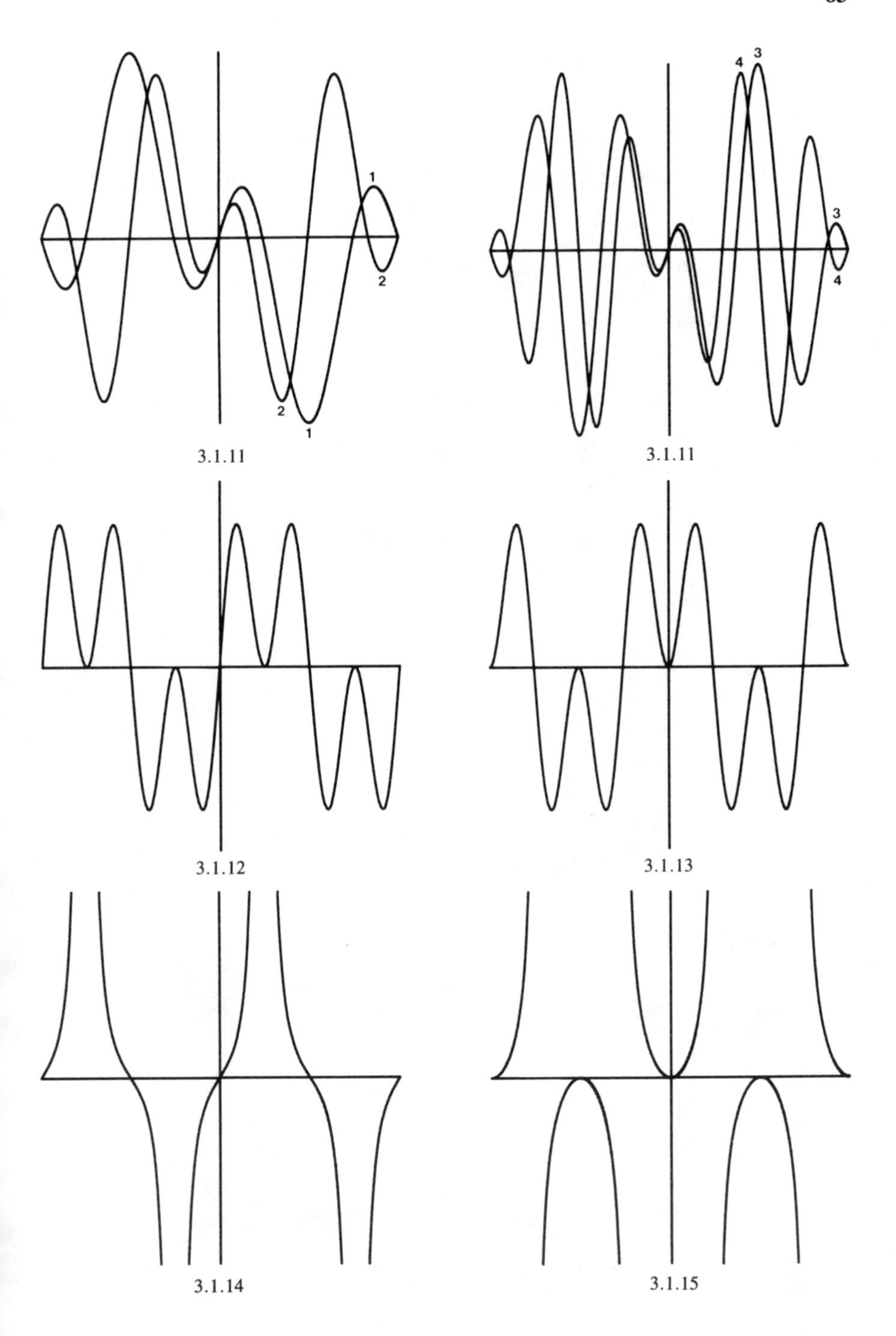

3.1.11

3.1.11

3.1.12

3.1.13

3.1.14

3.1.15

3.1.16. $y = 0.25 \cos(2\pi x)/\sin^2(2\pi x)$

3.1.17. $y = 0.25 \cos^2(2\pi x)/\sin(2\pi x)$

3.2. TRIGONOMETRIC FUNCTIONS WITH $1 \pm \sin^n(ax)$ AND $1 \pm \cos^m(bx)$

3.2.1. $y = 0.5/(1 + \cos(2\pi x))$

3.2.2. $y = 0.5/(1 - \cos(2\pi x))$

3.2.3. $y = 0.5 \sin(2\pi x)/[1 + \cos(2\pi x)]$

3.2.4. $y = 0.5 \sin(2\pi x)/[1 - \cos(2\pi x)]$

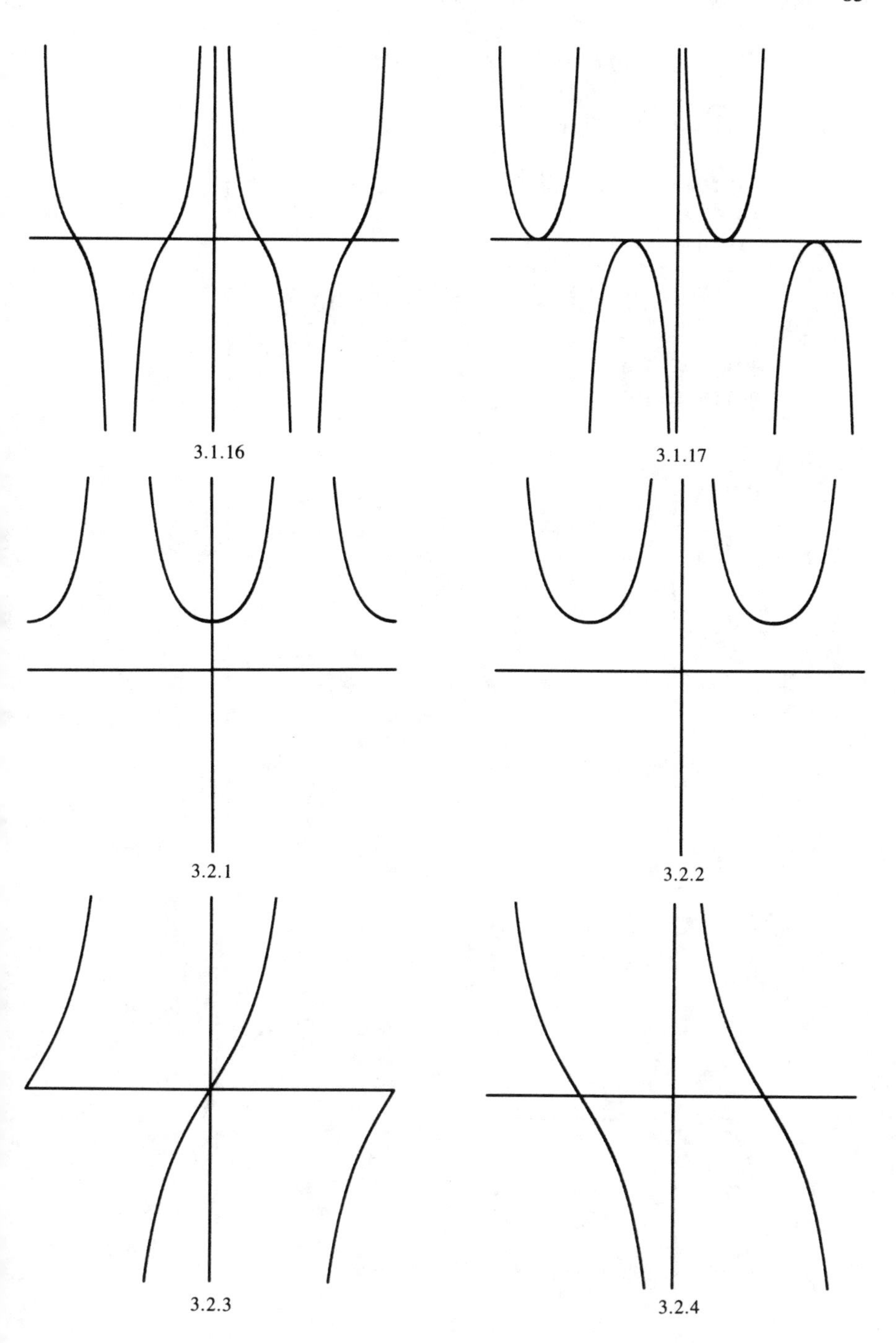

3.1.16

3.1.17

3.2.1

3.2.2

3.2.3

3.2.4

3.2.5. $y = 0.5 \cos(2\pi x)/[1 + \cos(2\pi x)]$

3.2.6. $y = 0.5 \cos(2\pi x)/[1 - \cos(2\pi x)]$

3.2.7. $y = 0.5/[1 + \cos(2\pi x)]^{1/2}$

3.2.8. $y = 0.5/[1 - \cos(2\pi x)]^{1/2}$

3.2.9. $y = 0.5/[a^2 + b^2\cos^2(2\pi x)]$
1. $a = 0.0, b = 1.0$
2. $a = 1.0, b = 1.0$
3. $a = 2.0, b = 1.0$

3.2.10. $y = 0.5/[a^2 - b^2\cos^2(2\pi x)]$
1. $a = 0.0, b = 1.0$
2. $a = 1.0, b = 1.0$
3. $a = 2.0, b = 1.0$

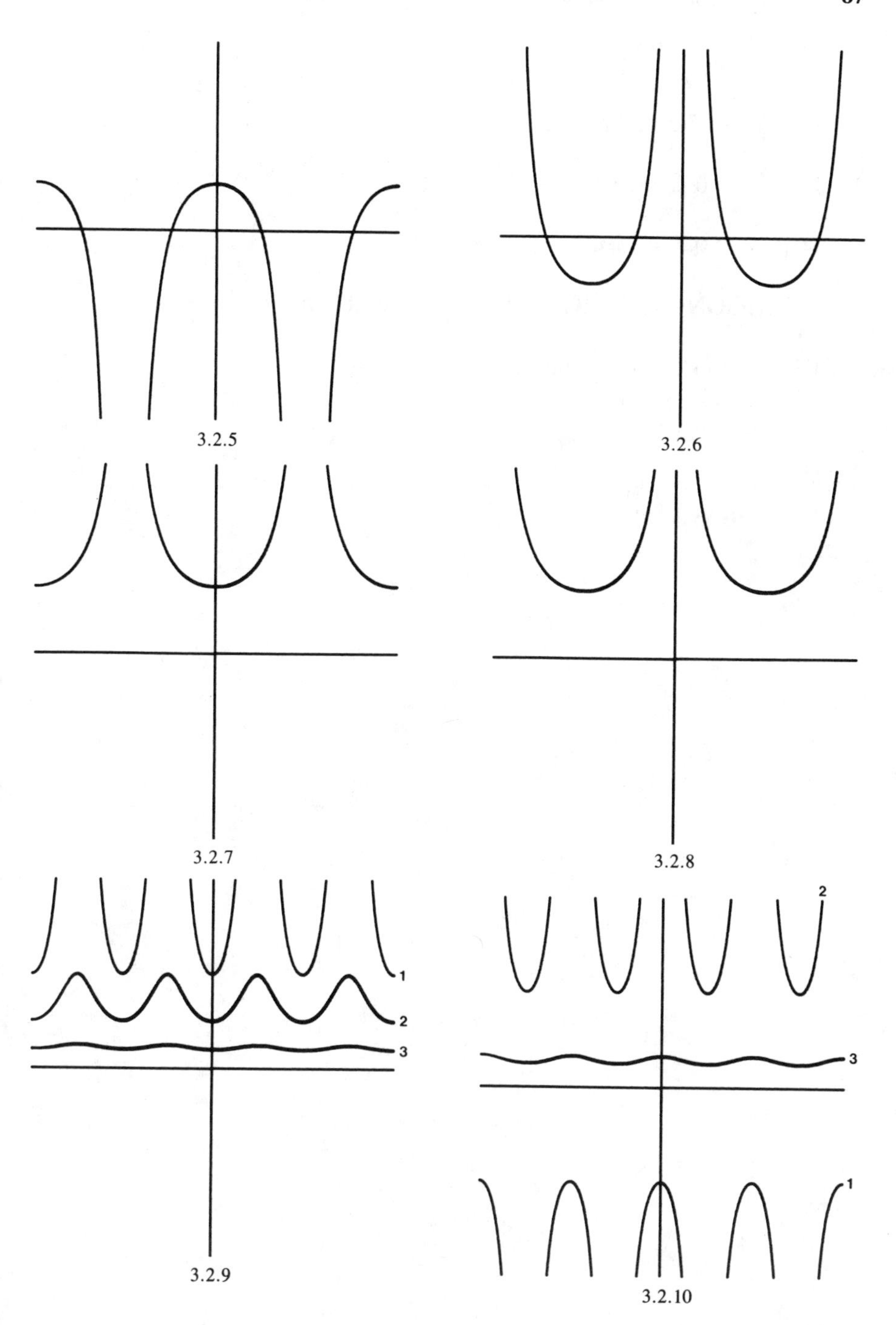

3.2.5

3.2.6

3.2.7

3.2.8

3.2.9

3.2.10

3.2.11. $y = \sin^2(2\pi x)/[1 + \cos^2(2\pi x)]$

3.2.12. $y = \cos^2(2\pi x)/[1 + \sin^2(2\pi x)]$

3.2.13. $y = 2.0 \sin^2(2\pi x)/[1 + \sin^2(2\pi x)]$

3.2.14. $y = 2.0 \cos^2(2\pi x)/[1 + \cos^2(2\pi x)]$

3.3. TRIGONOMETRIC FUNCTIONS WITH $a \cdot \sin^n(cx) + b \cdot \cos^m(cx)$

3.3.1. $y = a \cdot \cos(2\pi x) + b \cdot \sin(2\pi x)$
1. $a = 0.4$, $b = 0.8$
2. $a = 0.8$, $b = 0.4$

3.3.2. $y = 1/[a \cdot \cos(2\pi x) + b \cdot \sin(2\pi x)]$
1. $a = 1.0$, $b = 2.0$
2. $a = 1.0$, $b = 4.0$

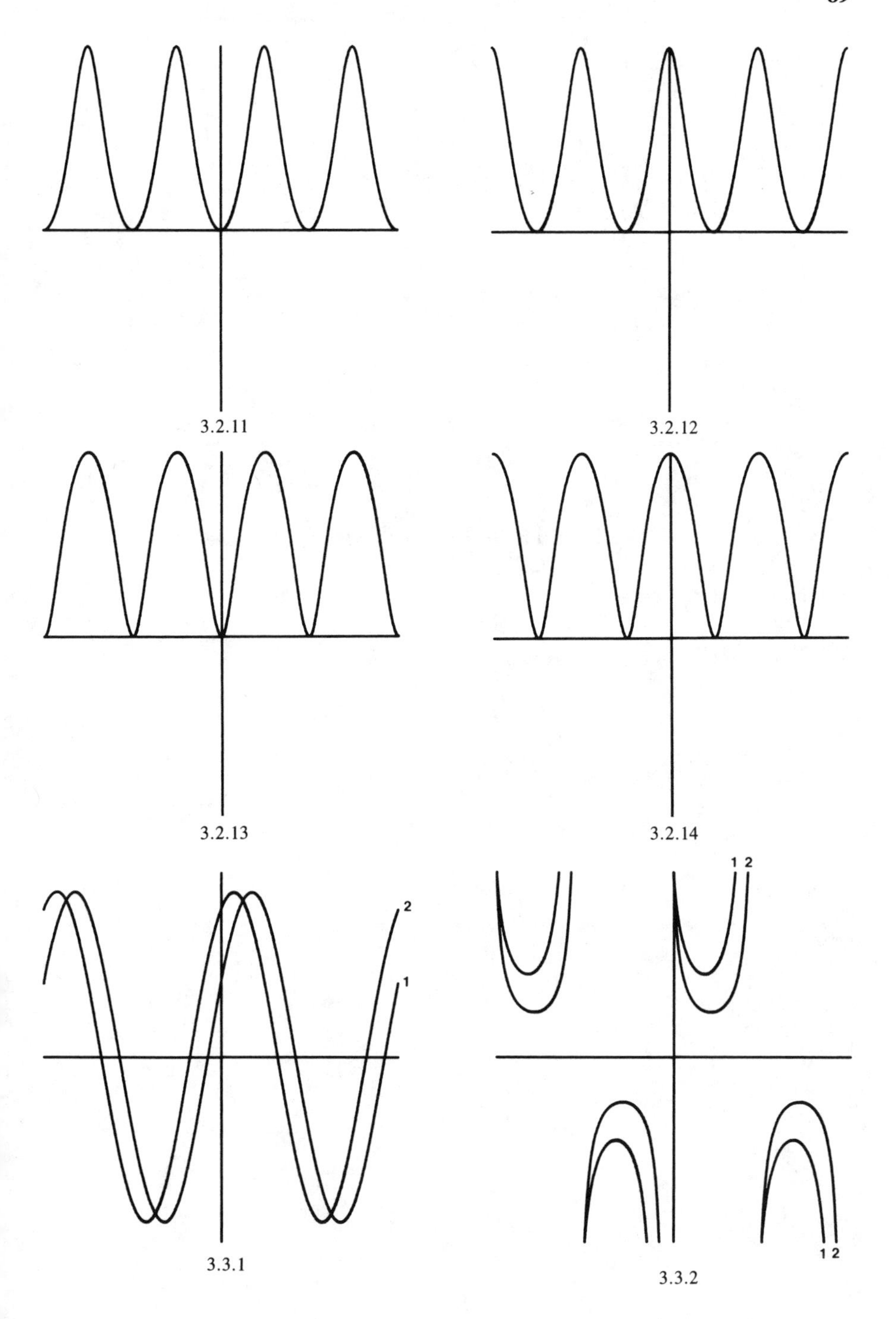

3.2.11

3.2.12

3.2.13

3.2.14

3.3.1

3.3.2

3.3.3. $y = a^2\cos^2(2\pi x) + b^2\sin^2(2\pi x)$
1. a = 0.25, b = 1.0
2. a = 0.50, b = 1.0

3.3.4. $y = 1/[a^2\cos^2(2\pi x) + b^2\sin^2(2\pi x)]$
1. a = 1.0, b = 2.0
2. a = 1.0, b = 4.0

3.3.5. $y = \sin(2\pi x)/[a\cdot\cos(2\pi x) + b\cdot\sin(2\pi x)]$
1. a = 1.0, b = 2.0
2. a = 1.0, b = 4.0

3.3.6. $y = \cos(2\pi x)/[a\cdot\cos(2\pi x) + b\cdot\sin(2\pi x)]$
1. a = 1.0, b = 2.0
2. a = 1.0, b = 4.0

3.4. INVERSE TRIGONOMETRIC FUNCTIONS

3.4.1. $y = (1/\pi)\arcsin(x)$

3.4.2. $y = (1/\pi)\arccos(x)$

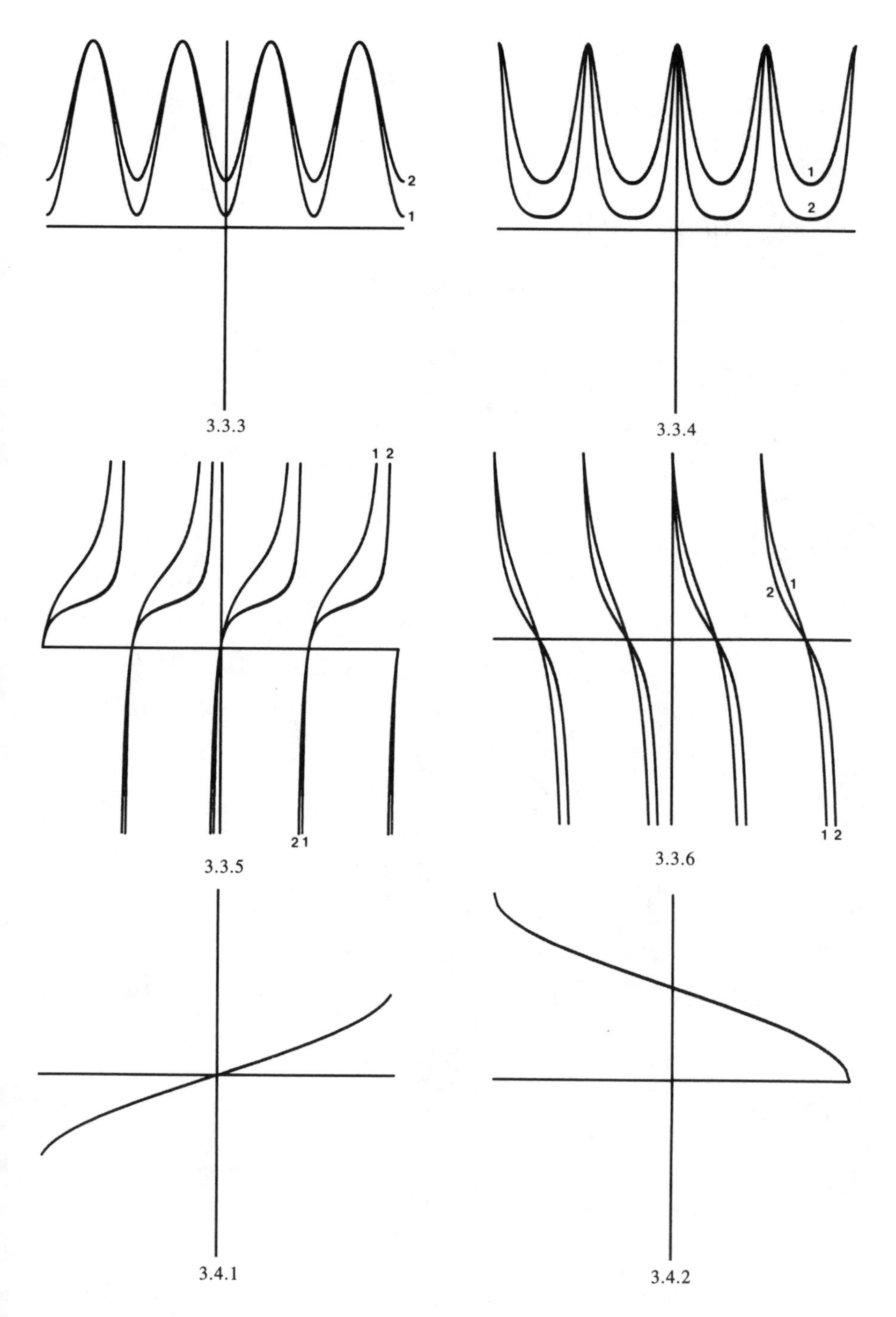

3.3.3

3.3.4

3.3.5

3.3.6

3.4.1

3.4.2

3.4.3. $y = (1/\pi)\arctan(10x)$

3.4.4. $y = (1/\pi)\text{arccot}(10x)$

3.4.5. $y = (1/\pi)\text{arcsec}(10x)$

3.4.6. $y = (1/\pi)\text{arccsc}(10x)$

3.5. LOGARITHMIC FUNCTIONS

3.5.1. $y = 0.25 \ln(10x)$

3.5.2. $y = 0.25 \ln(1/10x)$

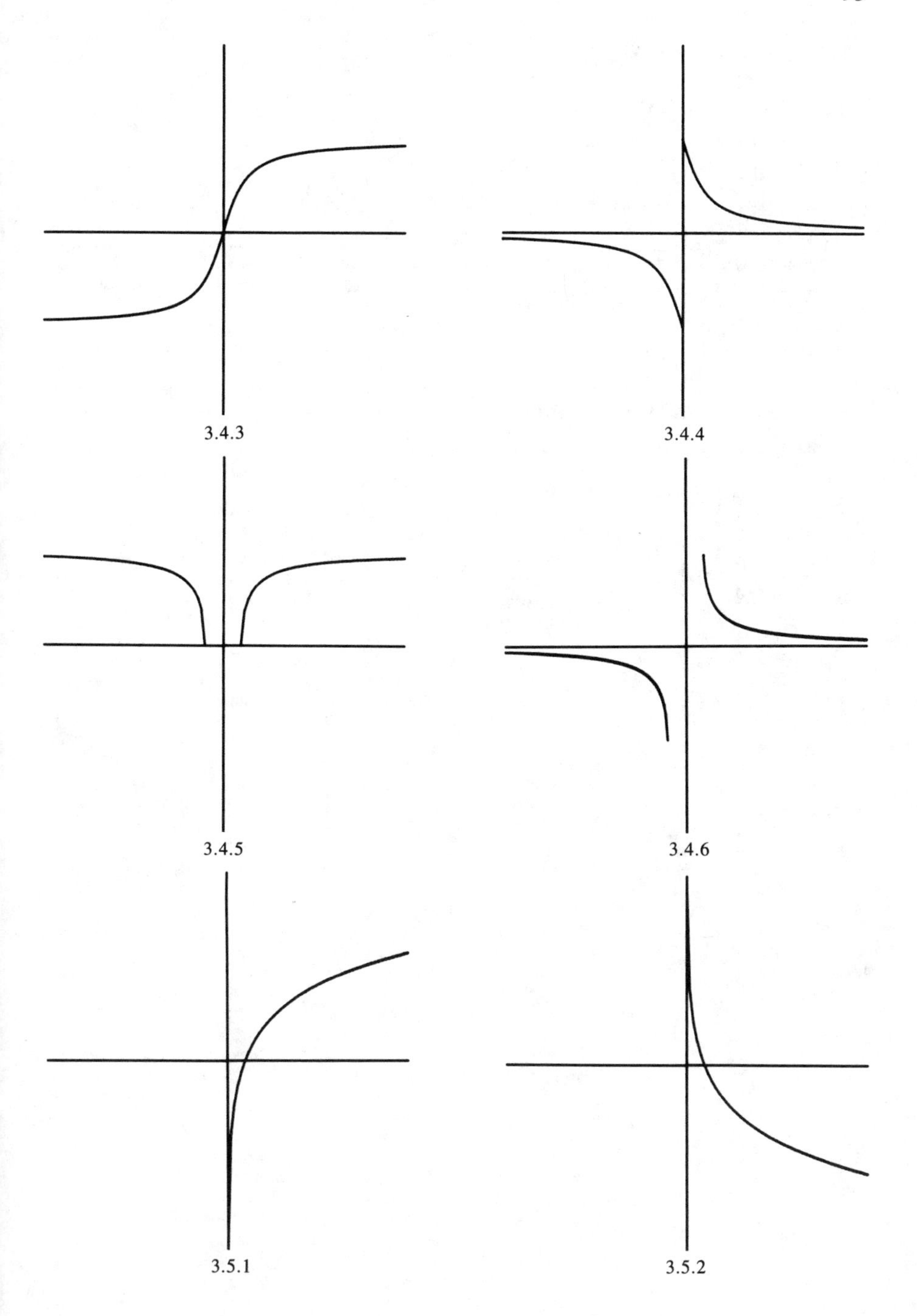
3.4.3
3.4.4
3.4.5
3.4.6
3.5.1
3.5.2

3.5.3. $y = 0.25/\ln(10x)$

3.5.4. $y = 0.5 \ln[(x + a)/(x - a)]$
1. $a = 0.1$
2. $a = 0.3$
3. $a = 0.5$

3.5.5. $y = 0.5 \ln(x^2 + a^2)$
1. $a = 0.5$
2. $a = 1.0$
3. $a = 2.0$

3.5.6. $y = 0.25 \ln[10(x^2 - a^2)]$
1. $a = 0.1$
2. $a = 0.3$
3. $a = 0.5$

3.5.7. $y = 0.5 \ln[x + (x^2 + a^2)^{1/2}]$
1. $a = 0.1$
2. $a = 0.3$
3. $a = 0.5$

3.5.8. $y = 0.5 \ln[x + (x^2 - a^2)^{1/2}]$
1. $a = 0.1$
2. $a = 0.3$
3. $a = 0.5$

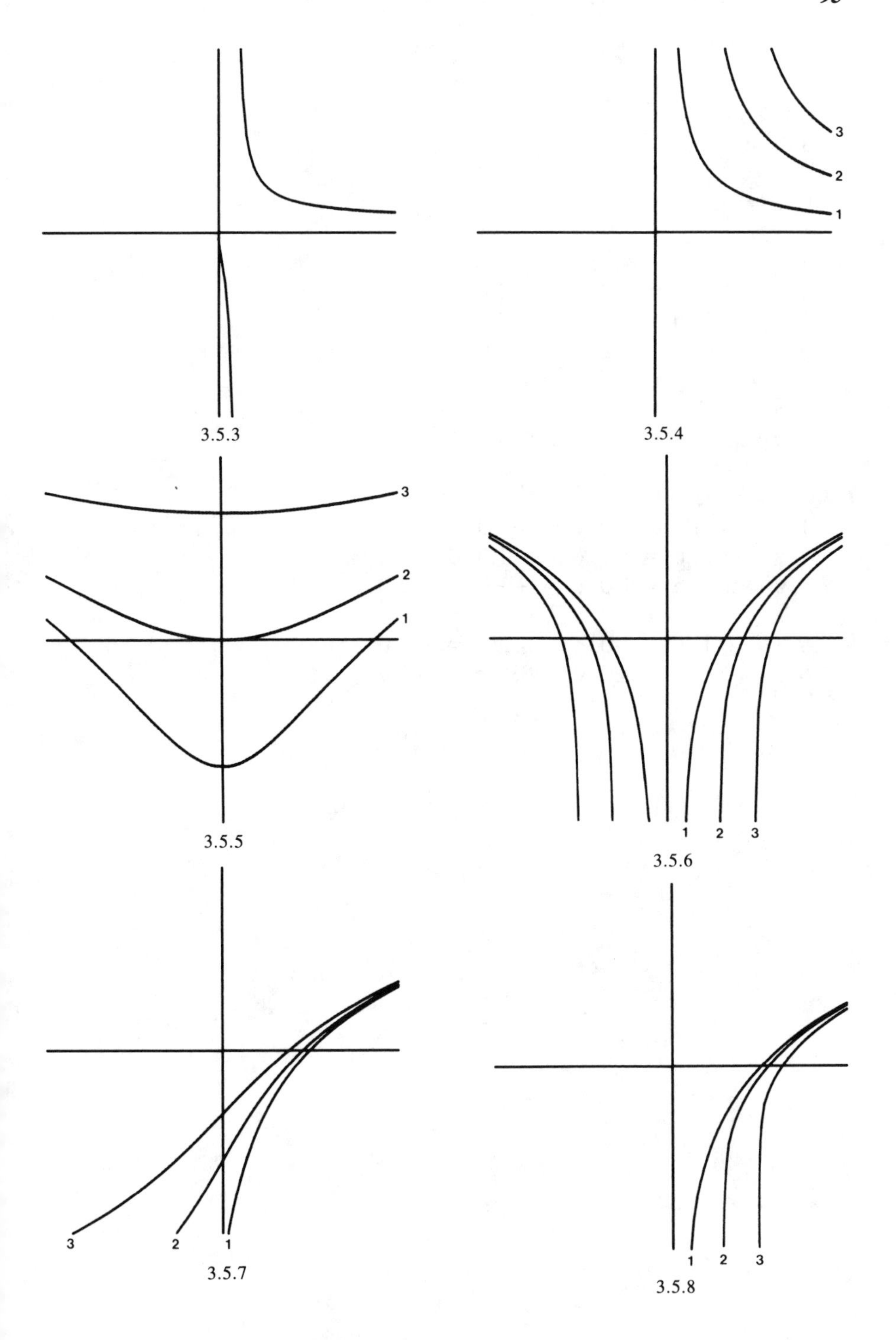

3.5.3

3.5.4

3.5.5

3.5.6

3.5.7

3.5.8

3.6. EXPONENTIAL FUNCTIONS

3.6.1. $y = 0.1\, e^{ax}$

1. $a = 1.0$
2. $a = 2.0$
3. $a = 3.0$

3.6.2. $y = 1/(a + be^{cx})$

1. $a = 1.0$, $b = 1.0$, $c = 2.0$
2. $a = 1.0$, $b = 1.0$, $c = 4.0$
3. $a = 1.0$, $b = 2.0$, $c = 4.0$
4. $a = -2.0$, $b = 2.0$, $c = 2.0$
5. $a = -2.0$, $b = 2.0$, $c = 4.0$
6. $a = -2.0$, $b = 4.0$, $c = 4.0$

3.6.3. $y = ae^{bx} + ce^{dx}$

1. $a = 1.0$, $b = 2.0$, $c = -1.0$, $d = 3.0$
2. $a = 1.0$, $b = 1.0$, $c = -1.0$, $d = 3.0$
3. $a = 1.0$, $b = 1.0$, $c = -1.0$, $d = 2.0$
4. $a = 0.1$, $b = 2.0$, $c = 0.1$, $d = -3.0$
5. $a = 0.1$, $b = 1.0$, $c = 0.1$, $d = -3.0$
6. $a = 0.1$, $b = 1.0$, $c = 0.1$, $d = -2.0$

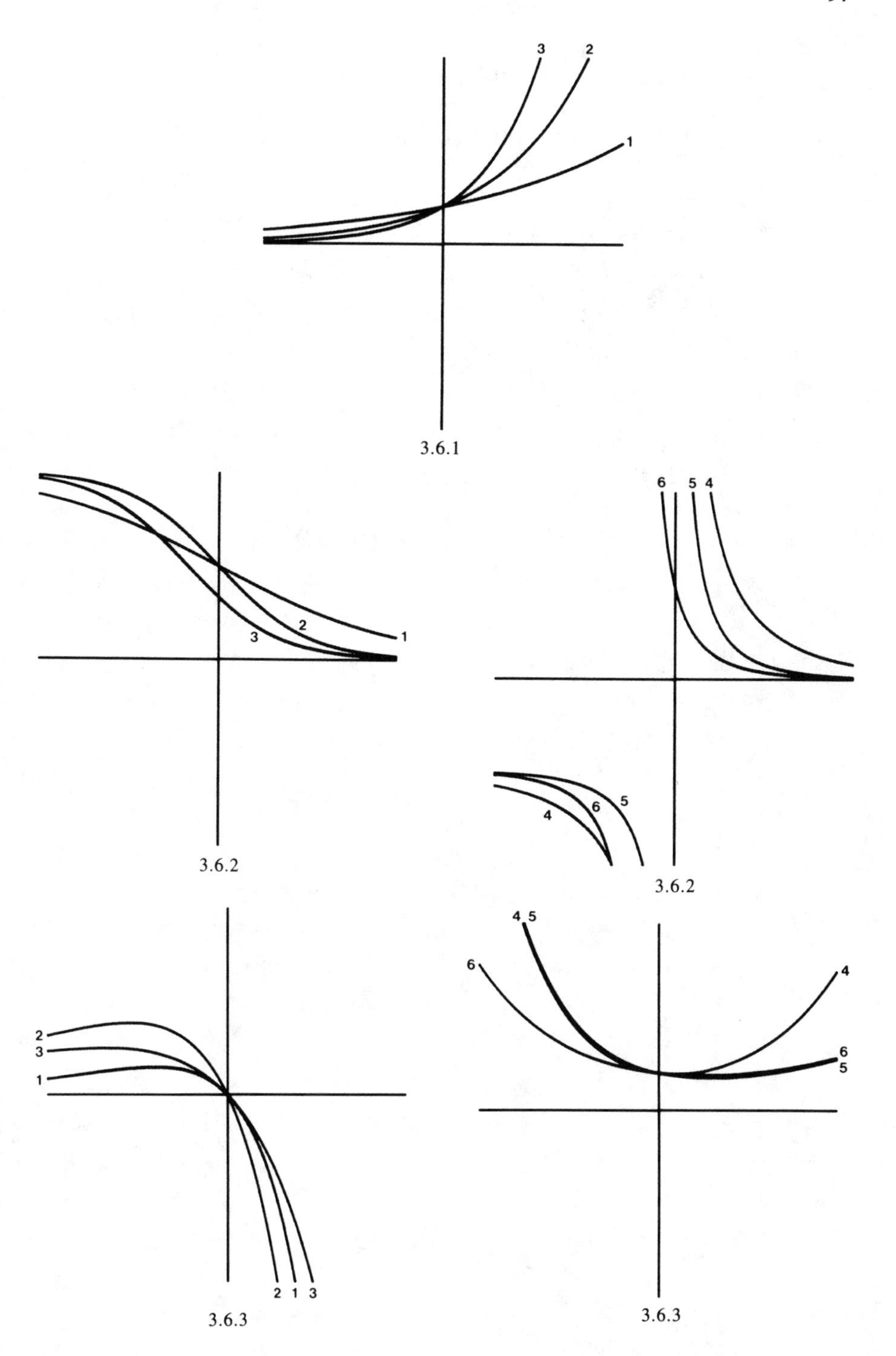
3
2
1
3.6.1
3
2
1
3.6.2
6
5
4
6
5
4
3.6.2
2
3
1
2
1
3
3.6.3
4
5
6
4
6
5
3.6.3

3.6.4. $y = 1/(ae^{bx} + ce^{dx})$

1. $a = 10.0, b = 2.0, c = -10.0, d = 3.0$
2. $a = 10.0, b = 1.0, c = -10.0, d = 3.0$
3. $a = 10.0, b = 1.0, c = -10.0, d = 2.0$
4. $a = 1.0, b = 2.0, c = 1.0, d = -3.0$
5. $a = 1.0, b = 1.0, c = 1.0, d = -3.0$
6. $a = 1.0, b = 1.0, c = 1.0, d = -2.0$

3.6.5. $y = c \cdot \exp(ax^2)$

1. $a = -1.0, c = 1.0$
2. $a = -2.0, c = 1.0$
3. $a = -3.0, c = 1.0$
4. $a = 1.0, c = 0.3$
5. $a = 2.0, c = 0.3$
6. $a = 3.0, c = 0.3$

3.7. HYPERBOLIC FUNCTIONS

3.7.1. $y = 0.1 \sinh(5x)$

3.7.2. $y = 0.1 \cosh(5x)$

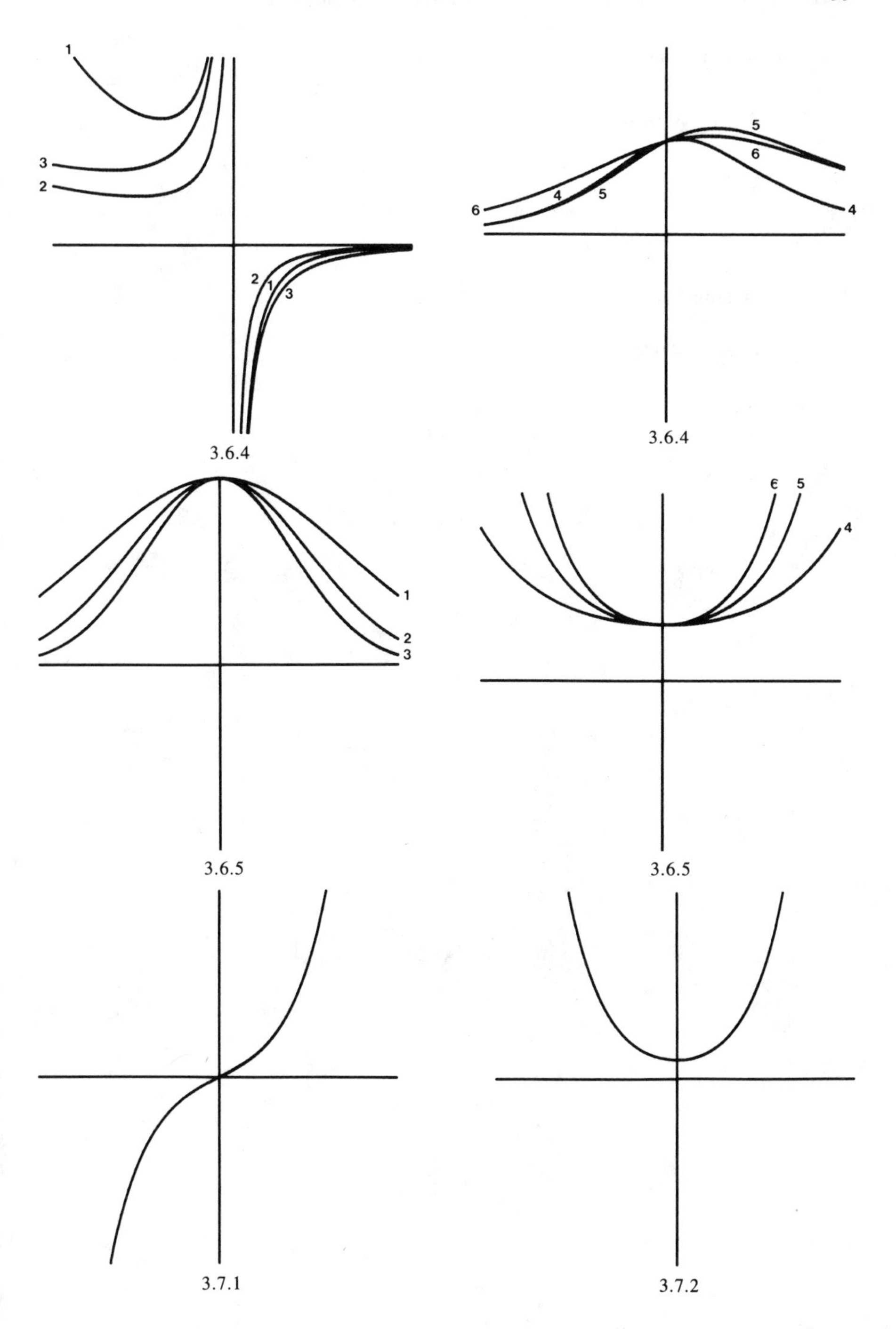

3.6.4

3.6.4

3.6.5

3.6.5

3.7.1

3.7.2

3.7.3. $y = \tanh(5x)$

3.7.4. $y = 0.1 \coth(5x)$

3.7.5. $y = \operatorname{sech}(5x)$

3.7.6. $y = 0.1 \operatorname{csch}(5x)$

3.7.7. $y = \sinh^2(x)$

3.7.8. $y = 0.5 \cosh^2(x)$

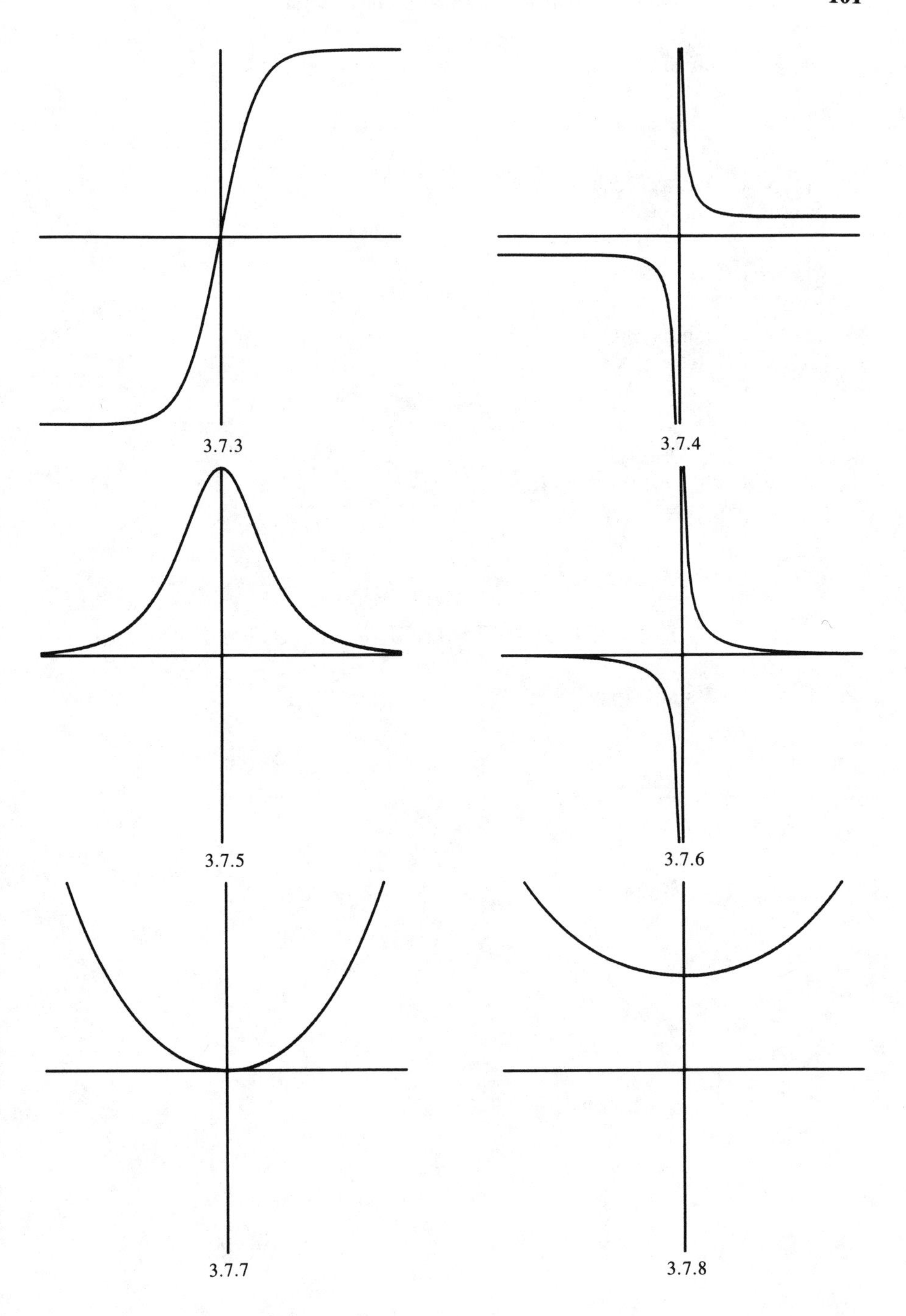
3.7.3
3.7.4
3.7.5
3.7.6
3.7.7
3.7.8

3.7.9. $y = \tanh^2(5x)$

3.7.10. $y = 0.25/[\sinh(x)\cdot\cosh(x)]$

3.7.11. $y = \sinh(ax)\cdot\cosh(bx)$
1. $a = 0.5$, $b = 0.75$
2. $a = 1.0$, $b = 0.75$
3. $a = 1.0$, $b = 1.50$

3.7.12. $y = \sinh(ax)\cdot\sinh(bx)$
1. $a = 1.0$, $b = 1.5$
2. $a = 1.0$, $b = 2.0$
3. $a = 1.0$, $b = 3.0$

3.7.13. $y = 0.5 \cosh(ax)\cdot\cosh(bx)$
1. $a = 1.0$, $b = 1.25$
2. $a = 1.0$, $b = 2.00$
3. $a = 1.0$, $b = 4.00$

3.8. INVERSE HYPERBOLIC FUNCTIONS

3.8.1. $y = 0.5 \sinh^{-1}(5x)$

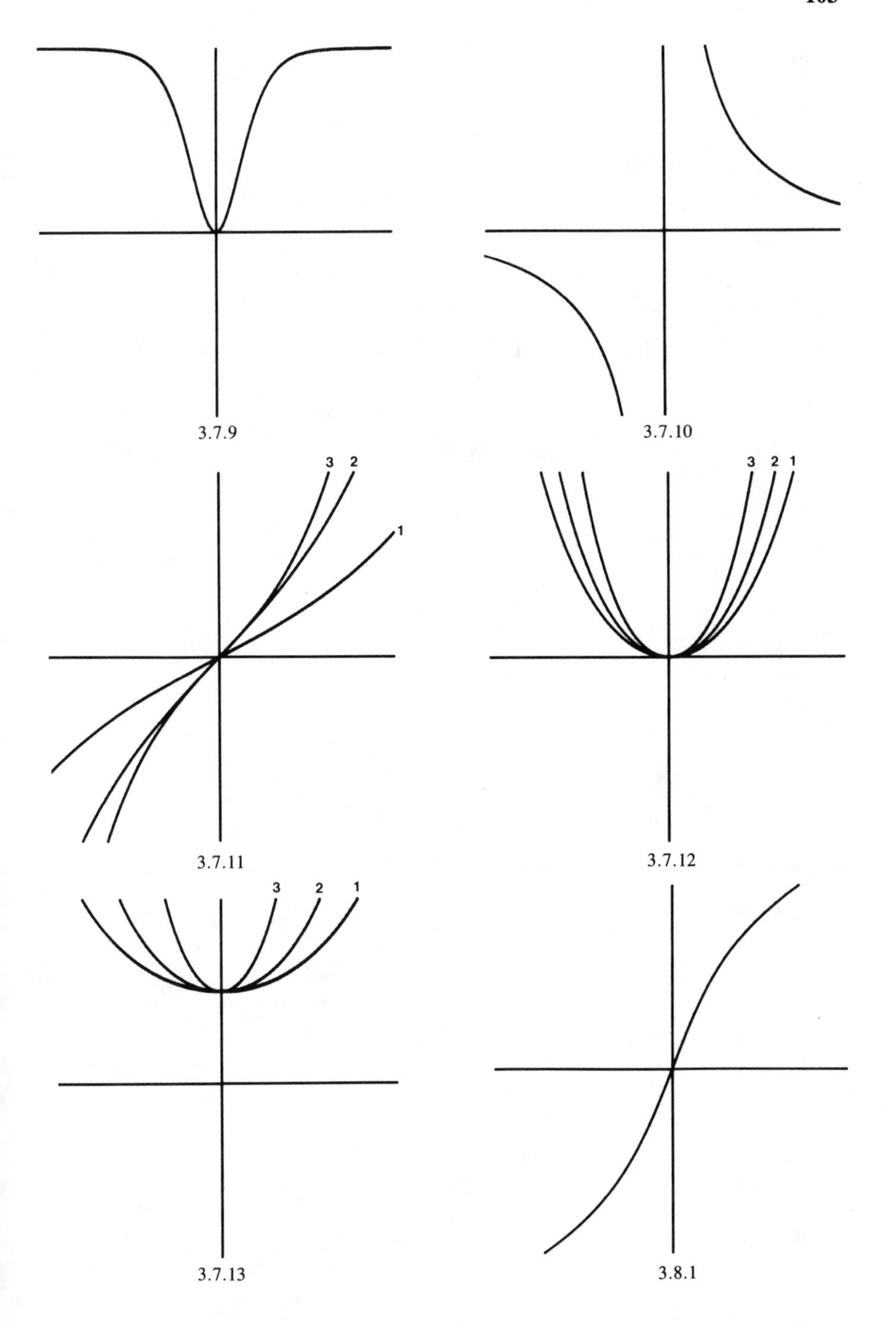

3.7.9

3.7.10

3.7.11

3.7.12

3.7.13

3.8.1

3.8.2. $y = 0.5 \cosh^{-1}(5x)$

3.8.3. $y = 0.2 \tanh^{-1}(x)$

3.8.4. $y = \coth^{-1}(5x)$

3.8.5. $y = 0.2 \operatorname{sech}^{-1}(x)$

3.8.6. $y = 0.2 \operatorname{csch}^{-1}(x)$

3.9. TRIGONOMETRIC AND EXPONENTIAL FUNCTIONS COMBINED

3.9.1. $y = e^{ax}\sin(2\pi bx)$

1. $a = -1.0$, $b = 4.0$
2. $a = -2.0$, $b = 4.0$

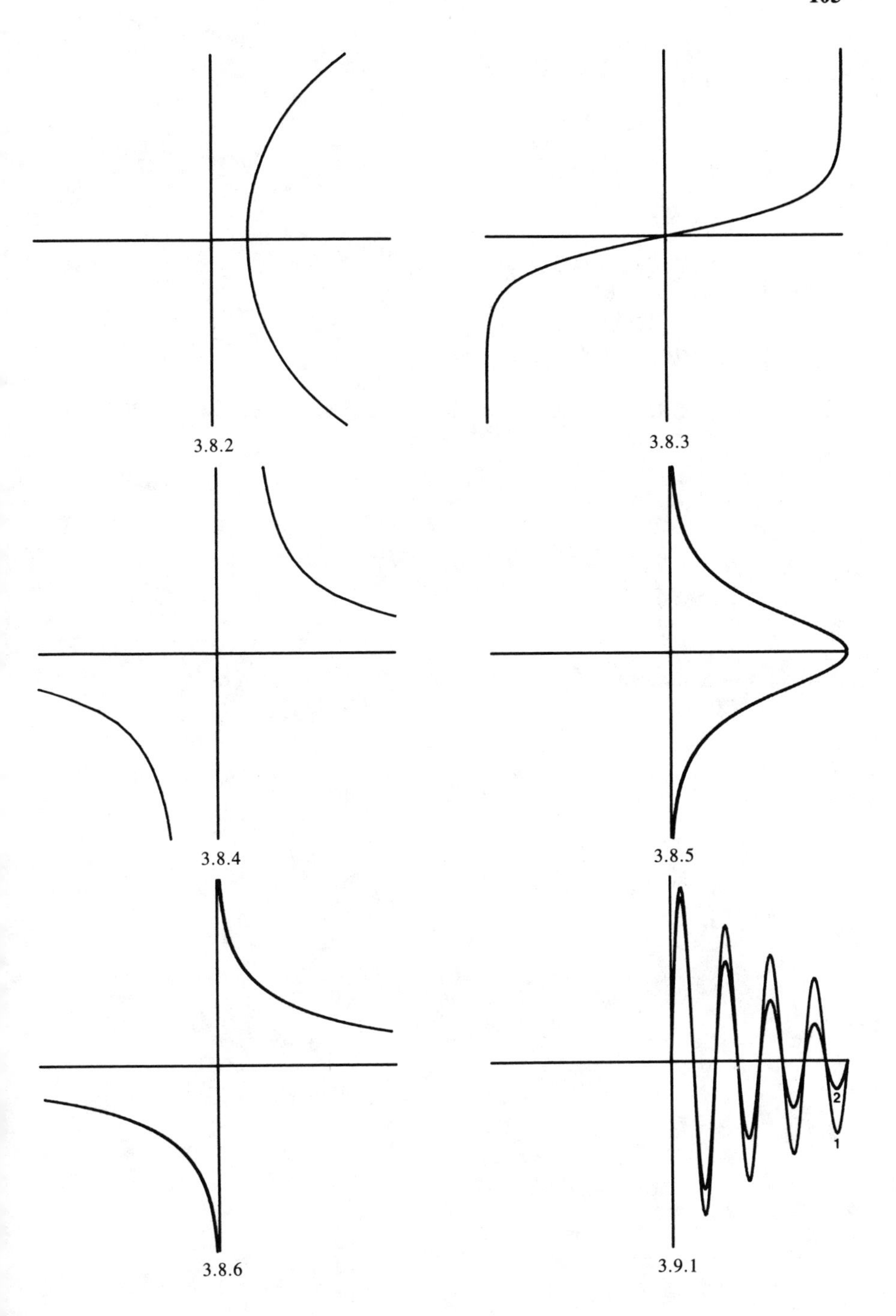

3.8.2

3.8.3

3.8.4

3.8.5

3.8.6

3.9.1

3.9.2. $y = e^{ax}\cos(2\pi bx)$
1. $a = -1.0$, $b = 4.0$
2. $a = -2.0$, $b = 4.0$

3.9.3. $y = 0.5\ e^{ax}/\sin(2\pi bx)$
1. $a = -1.0$, $b = 4.0$
2. $a = -2.0$, $b = 4.0$

3.9.4. $y = 0.5\ e^{ax}/\cos(2\pi bx)$
1. $a = -1.0$, $b = 4.0$
2. $a = -2.0$, $b = 4.0$

3.10. TRIGONOMETRIC FUNCTIONS COMBINED WITH POWERS OF x

3.10.1. $y = x \cdot \sin(2\pi ax)$
1. $a = 4.0$

3.10.2. $y = x \cdot \cos(2\pi ax)$
1. $a = 4.0$

3.10.3. $y = x/\sin(2\pi ax)$
1. $a = 4.0$

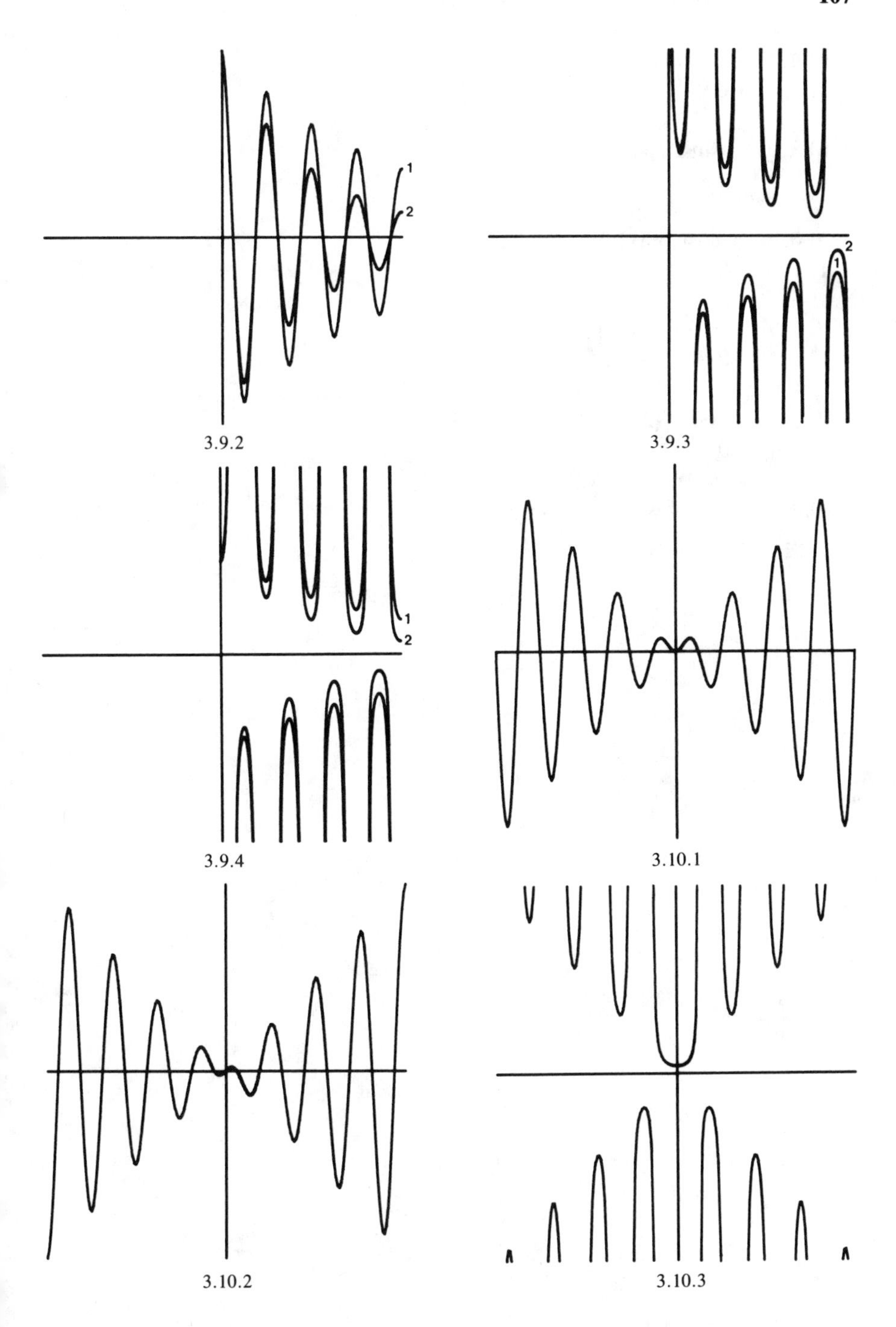

3.9.2

3.9.3

3.9.4

3.10.1

3.10.2

3.10.3

3.10.4. $y = x/\cos(2\pi ax)$
1. $a = 4.0$

3.10.5. $y = \sin(2\pi ax)/2\pi ax$
1. $a = 4.0$

3.10.6. $y = \cos(2\pi ax)/2\pi ax$
1. $a = 4.0$

3.10.7. $y = x \cdot \sin^2(2\pi ax)$
1. $a = 4.0$

3.10.8. $y = x \cdot \cos^2(2\pi ax)$
1. $a = 4.0$

3.10.9. $y = 0.025 \sin(2\pi ax)/x^2$
1. $a = 4.0$

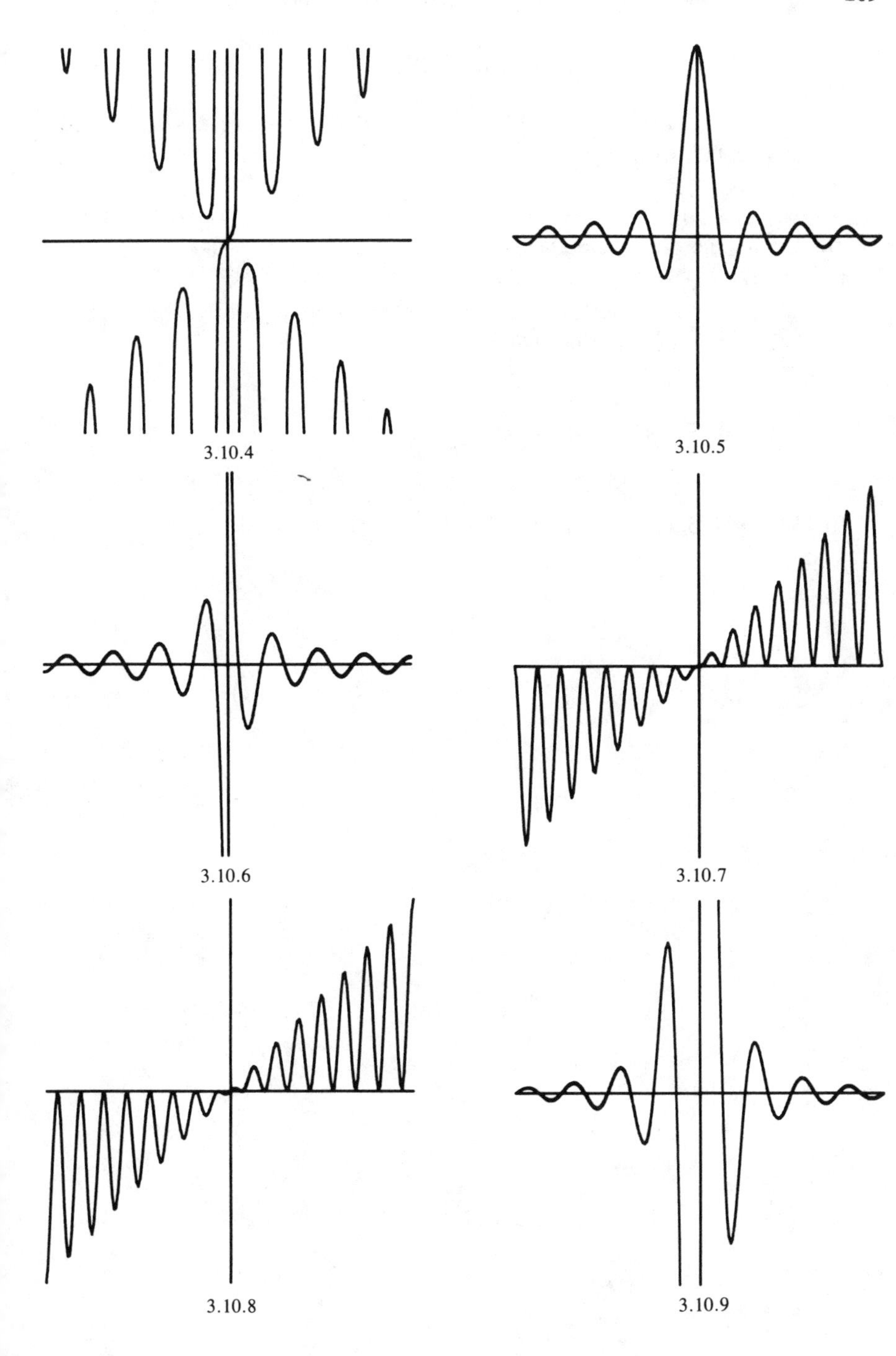

3.10.4

3.10.5

3.10.6

3.10.7

3.10.8

3.10.9

3.10.10. $y = 0.01 \cos(2\pi ax)/x^2$
1. $a = 4.0$

3.10.11. $y = 0.5\, x/\sin^2(2\pi ax)$
1. $a = 4.0$

3.10.12. $y = 0.5\, x/\cos^2(2\pi ax)$
1. $a = 4.0$

3.10.13. $y = 0.5\, x/[1 + \sin(2\pi ax)]$
1. $a = 4.0$

3.10.14. $y = 0.5\, x/[1 + \cos(2\pi ax)]$
1. $a = 4.0$

3.10.15. $y = 0.5\, x/[1 - \sin(2\pi ax)]$
1. $a = 4.0$

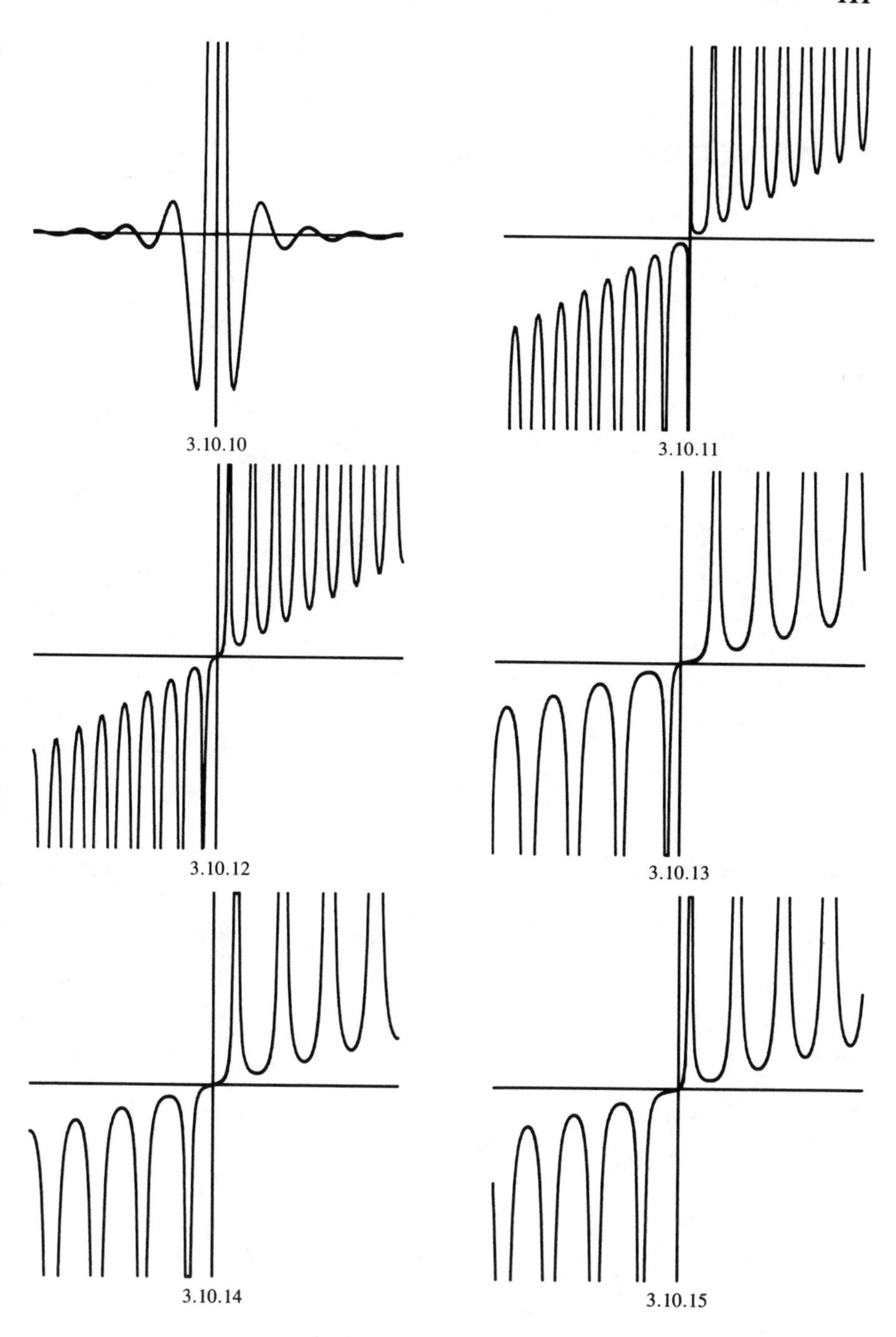

3.10.10

3.10.11

3.10.12

3.10.13

3.10.14

3.10.15

3.10.16. $y = 0.5\, x/[1 - \cos(2\pi ax)]$
1. $a = 4.0$

3.11. LOGARITHMIC FUNCTIONS COMBINED WITH POWERS OF x

3.11.1. $y = x \cdot \ln(ax)$
1. $a = 1.0$
2. $a = 2.0$
3. $a = 4.0$

3.11.2. $y = x^2 \ln(ax)$
1. $a = 1.0$
2. $a = 2.0$
3. $a = 4.0$

3.11.3. $y = 0.05/[x \cdot \ln(ax)]$
1. $a = 1.0$
2. $a = 2.0$
3. $a = 4.0$

3.11.4. $y = 0.005/[x^2 \ln(ax)]$
1. $a = 1.0$
2. $a = 2.0$
3. $a = 4.0$

3.11.5. $y = 0.1 \ln(ax)/x$
1. $a = 1.0$
2. $a = 3.0$
3. $a = 9.0$

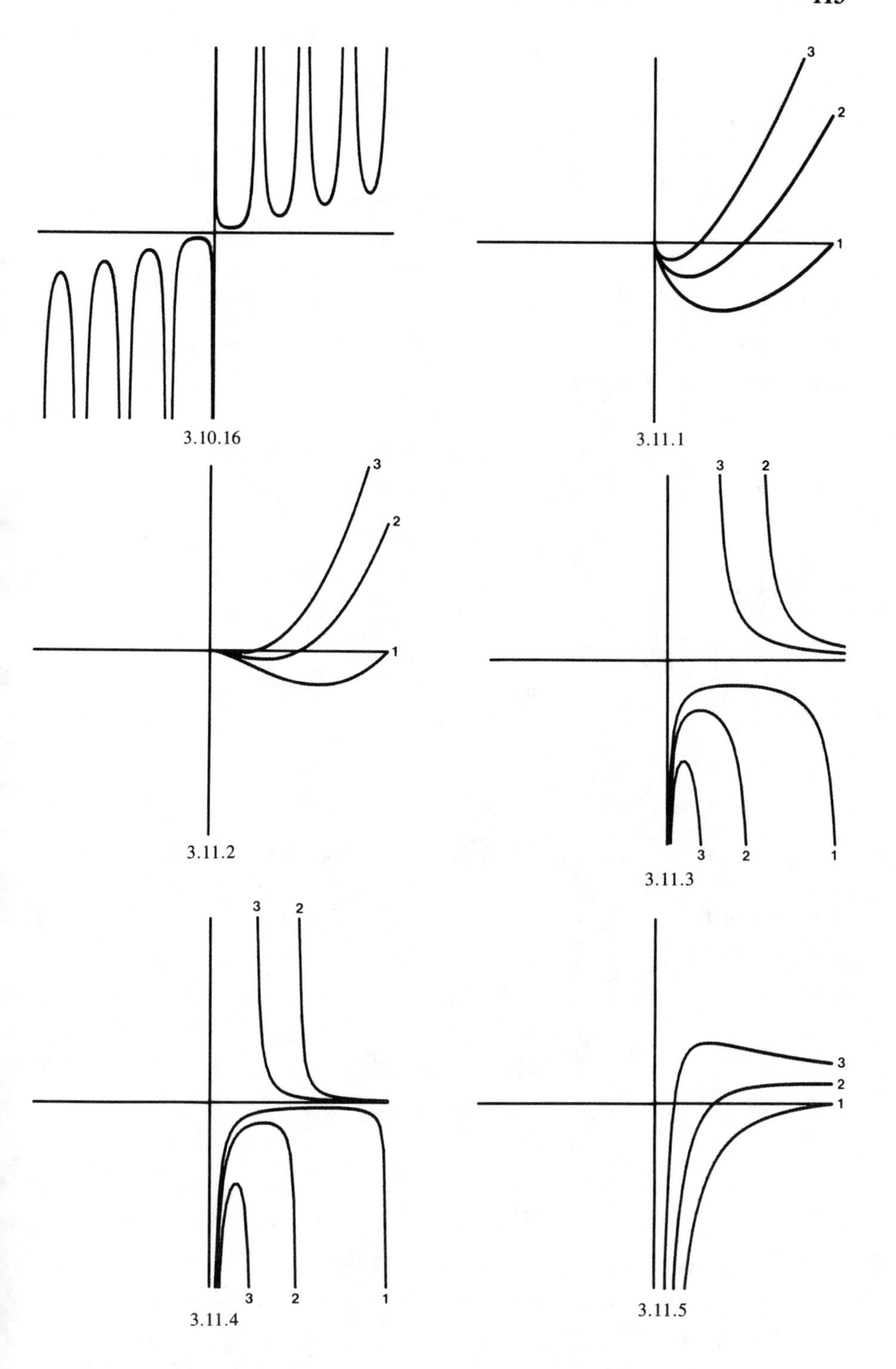

3.10.16

3.11.1

3.11.2

3.11.3

3.11.4

3.11.5

3.11.6. $y = 0.5\, x/\ln(ax)$
1. a = 1.0
2. a = 3.0
3. a = 9.0

3.11.7. $y = x \cdot \ln(ax + b)$
1. a = 1.0, b = 2.0
2. a = 4.0, b = 2.0
3. a = 4.0, b = −1.0
4. a = 4.0, b = −2.0

3.11.8. $y = 0.1 \ln(ax + b)/x$
1. a = 1.0, b = 2.0
2. a = 4.0, b = 2.0
3. a = 4.0, b = −1.0
4. a = 4.0, b = −2.0

3.11.9. $y = x \cdot \ln(x^2 + a^2)$
1. a = 0.0
2. a = 0.5
3. a = 1.0

3.11.10. $y = 0.5\, x \cdot \ln(x^2 - a^2)$
1. a = 0.1
2. a = 0.3
3. a = 0.5

3.12. EXPONENTIAL FUNCTIONS COMBINED WITH POWERS OF x

3.12.1. $y = 0.5\, xe^{ax}$
1. a = 0.5
2. a = 1.0
3. a = 2.0

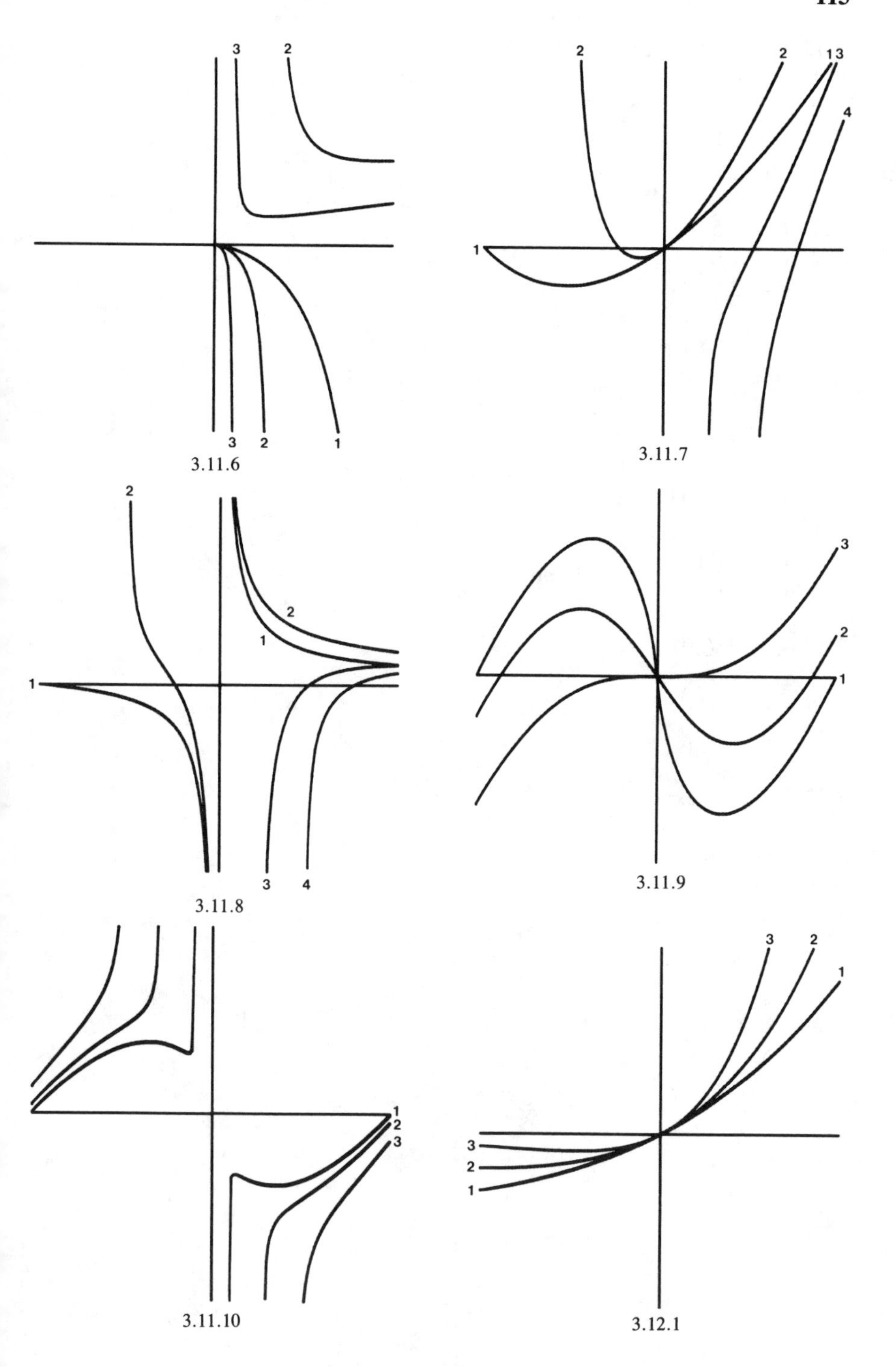

3.11.6

3.11.7

3.11.8

3.11.9

3.11.10

3.12.1

3.12.2. $y = 0.5\ x^2e^{ax}$
1. $a = 0.5$
2. $a = 1.0$
3. $a = 2.0$

3.12.3. $y = 0.5\ x^3e^{ax}$
1. $a = 0.5$
2. $a = 1.0$
3. $a = 2.0$

3.12.4. $y = 0.1\ e^{ax}/x$
1. $a = 1.0$
2. $a = 2.0$
3. $a = 3.0$

3.12.5. $y = 0.03\ e^{ax}/x^2$
1. $a = 2.0$
2. $a = 3.0$
3. $a = 4.0$

3.12.6. $y = 0.01\ e^{ax}/x^3$
1. $a = 3.0$
2. $a = 4.0$
3. $a = 5.0$

3.12.7. $y = x \cdot \exp(-ax^2)$
1. $a = 1.0$
2. $a = 2.0$
3. $a = 3.0$

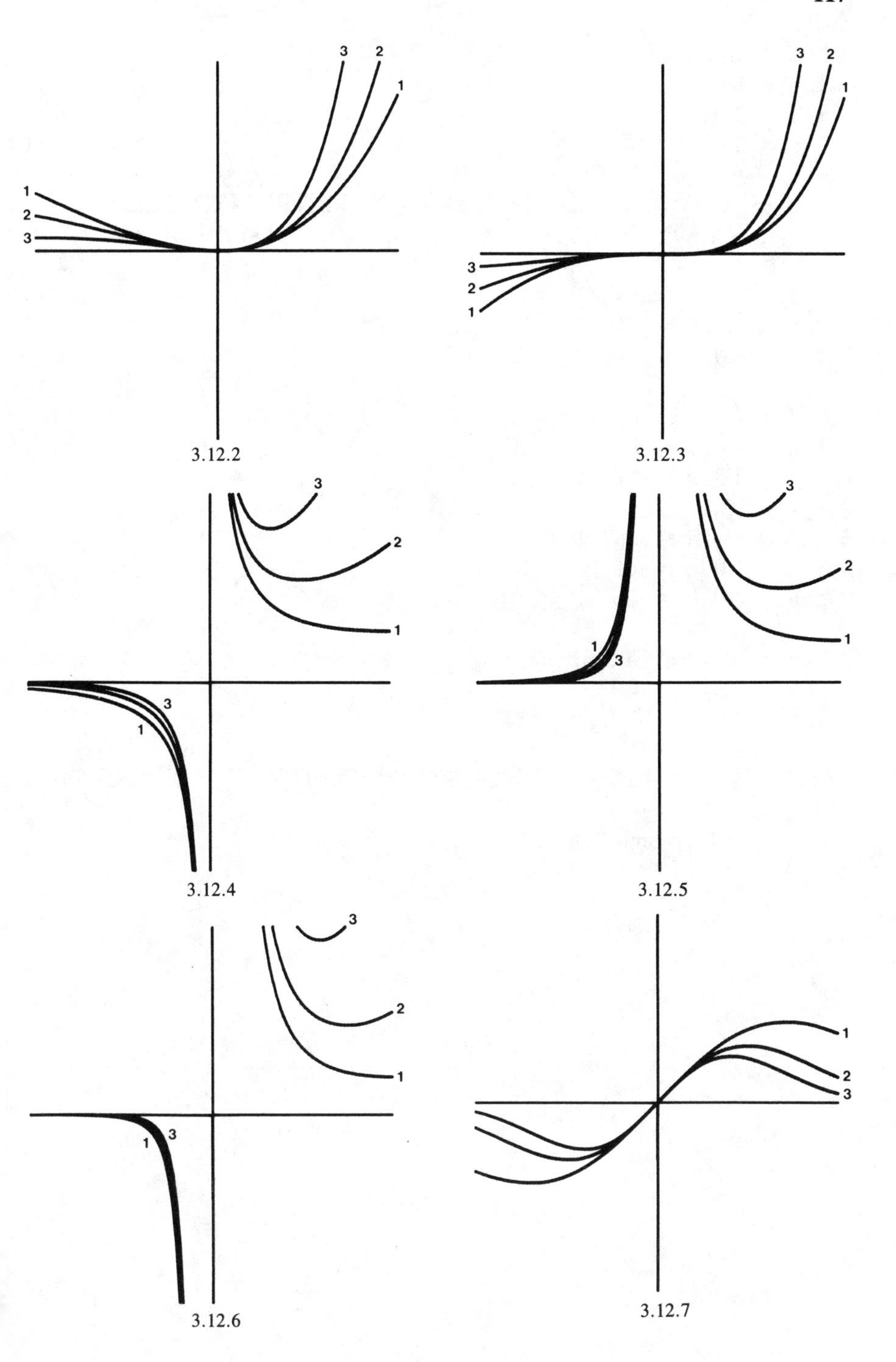

3.12.2

3.12.3

3.12.4

3.12.5

3.12.6

3.12.7

3.12.8. $y = 2.0\, x^2 \exp(-ax^2)$
1. $a = 1.0$
2. $a = 2.0$
3. $a = 3.0$

3.13. COMBINATIONS OF TRIGONOMETRIC FUNCTIONS, EXPONENTIAL FUNCTIONS, AND POWERS OF x

3.13.1. $y = 0.15\, xe^{ax}\sin(2\pi bx)$
1. $a = 1.0$, $b = 4.0$
2. $a = 2.0$, $b = 4.0$

3.13.2. $y = 0.15\, xe^{ax}\cos(2\pi bx)$
1. $a = 1.0$, $b = 4.0$
2. $a = 2.0$, $b = 4.0$

3.13.3. $y = 0.1\, e^{ax}\sin(2\pi bx)/x$
1. $a = 1.0$, $b = 4.0$
2. $a = 2.0$, $b = 4.0$

3.13.4. $y = 0.1\, e^{ax}\cos(2\pi bx)/x$
1. $a = 1.0$, $b = 4.0$
2. $a = 2.0$, $b = 4.0$

3.14. HYPERBOLIC FUNCTIONS COMBINED WITH POWERS OF x

3.14.1. $y = 0.1\, x \cdot \sinh(5x)$

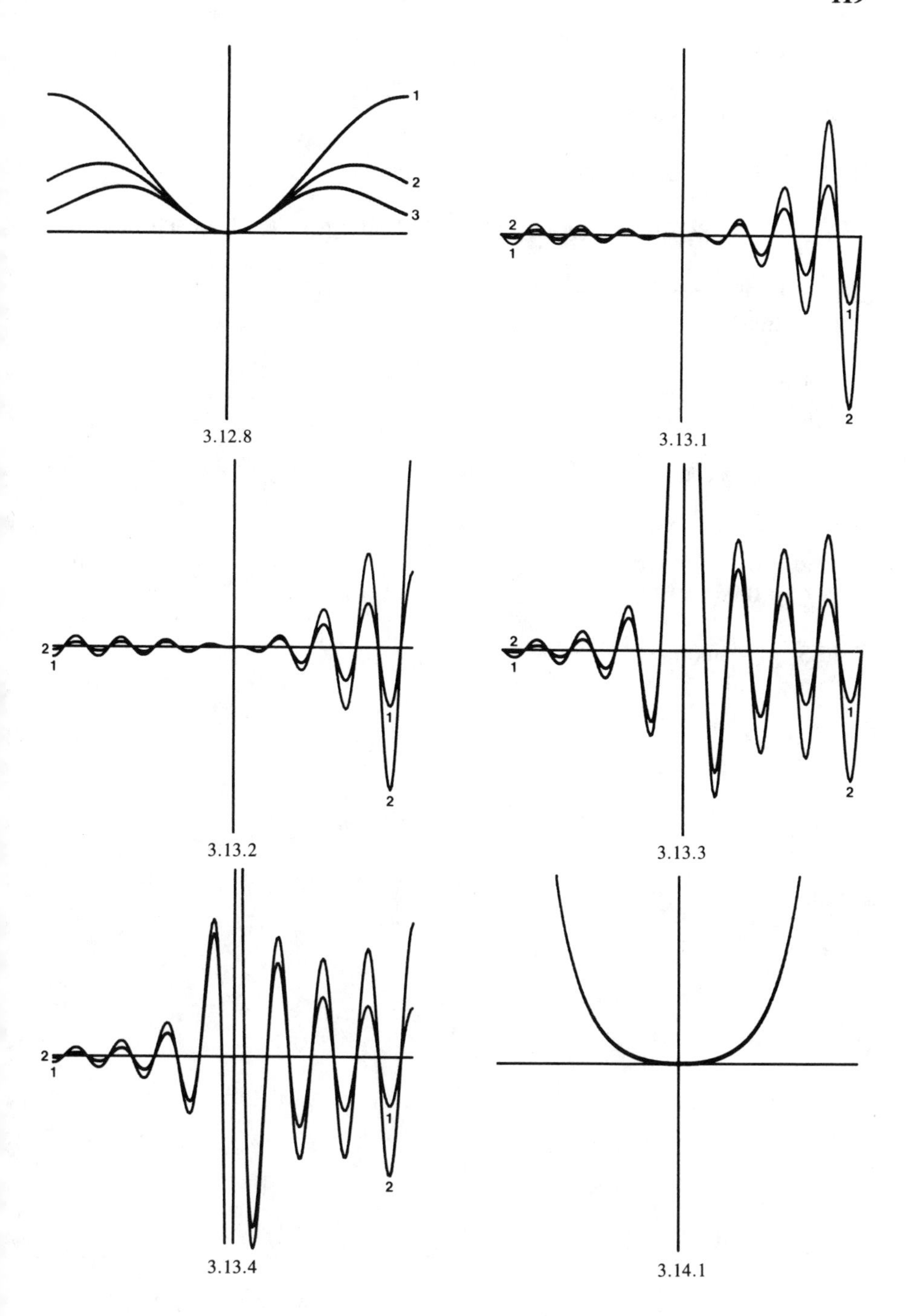

3.12.8

3.13.1

3.13.2

3.13.3

3.13.4

3.14.1

3.14.2. $y = 0.1\, x \cdot \cosh(5x)$

3.14.3. $y = 0.02 \sinh(5x)/x$

3.14.4. $y = 0.02 \cosh(5x)/x$

3.15. MISCELLANEOUS TRANSCENDENTAL FUNCTIONS

3.15.1. $y = a \cdot \cosh^{-1}(a/x) - (a^2 - x^2)^{1/2}$
"Tractrix"

1. $a = 1.0$
2. $a = 0.5$

3.15.2. $y = x \cdot \cot(\pi x/2a)$
"Quadratrix of Hippias"

1. $a = 0.25$
2. $a = 0.35$

3.15.3. $y = \sin(a\pi/x)$
1. $a = 1.0$

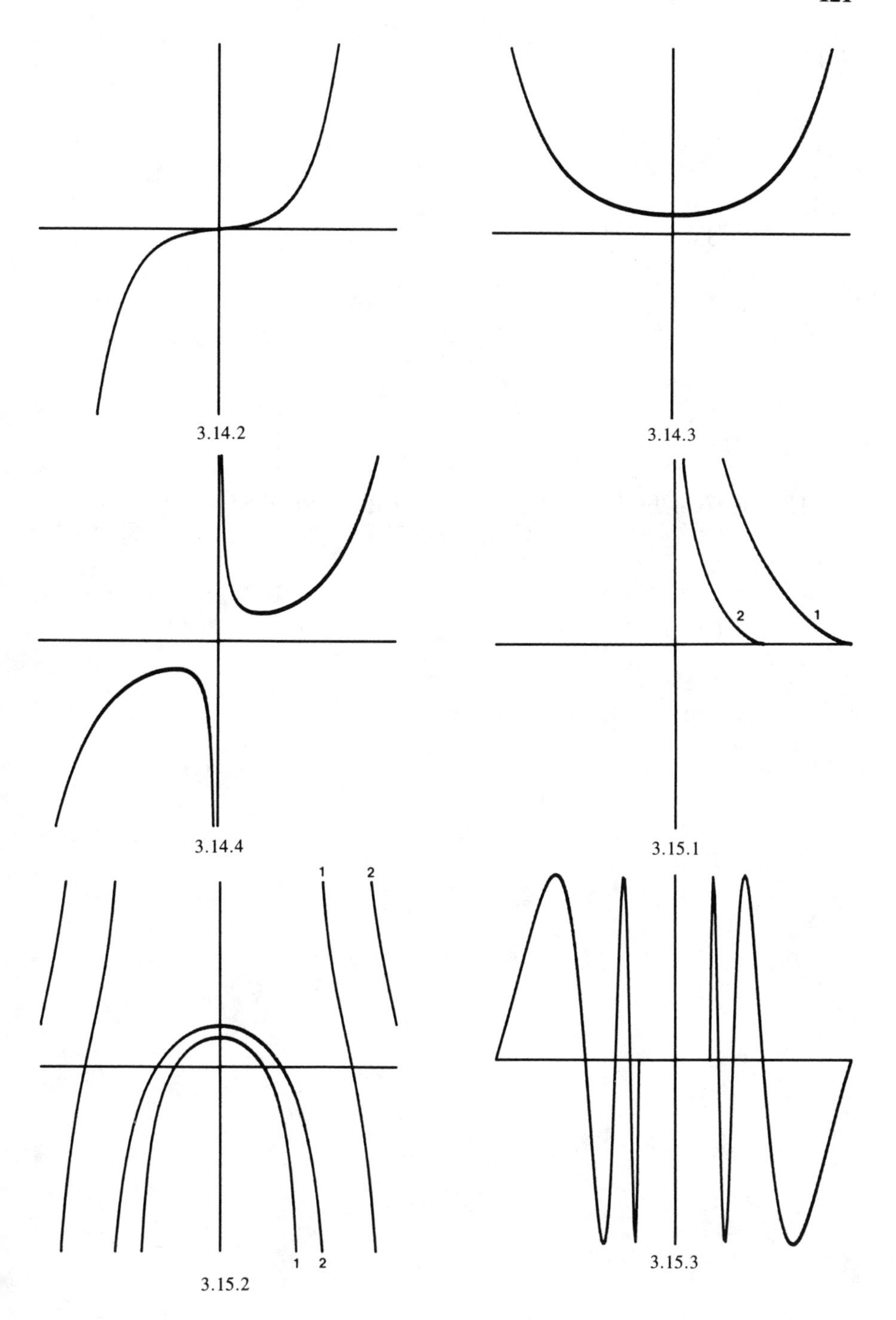

3.14.2

3.14.3

3.14.4

3.15.1

3.15.2

3.15.3

3.15.4. $y = \cos(a\pi/x)$
1. $a = 1.0$

3.15.5. $y = 1 - e^{ax}$
1. $a = -1.0$
2. $a = -3.0$
3. $a = -5.0$

3.15.6. $y = c \cdot \arctan(e^{ax}) - b$
Special case: $a = 1$, $b = \pi/2$, $c = 2$ gives "Gudermannian function"

1. $a = 1.0$, $b = \pi/4$, $c = 1.0$
2. $a = 3.0$, $b = \pi/4$, $c = 1.0$
3. $a = 10.0$, $b = \pi/4$, $c = 1.0$

3.16. TRANSCENDENTAL FUNCTIONS EXPRESSIBLE IN POLAR COORDINATES

3.16.1. $r = ce^{a\theta}$ $\ln[(x^2 + y^2)/c^2]/2 - a \cdot \arctan(y/x) = 0$
"Logarithmic spiral" (also called "equiangular spiral" or "logistique")

1. $a = 0.1$, $c = 0.10$; $0 < \theta < 15\pi/2$
2. $a = 0.2$, $c = 0.01$; $0 < \theta < 15\pi/2$

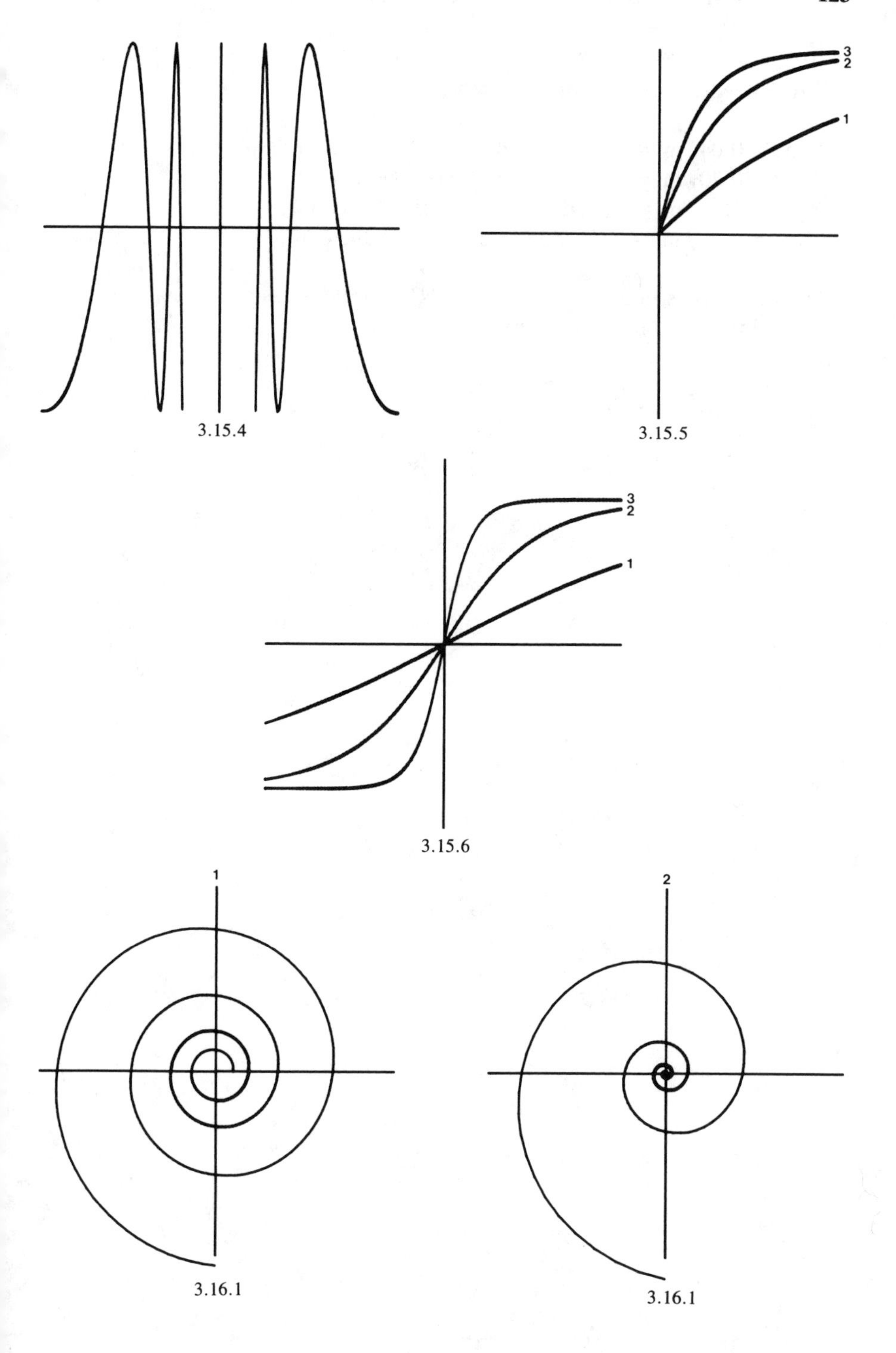

3.15.4

3.15.5

3.15.6

3.16.1

3.16.1

3.16.2. $r = a\,\theta^{1/m}$ $(x^2 + y^2)^{m/2} - a^m \arctan(y/x) = 0$

"Archimedean spirals" ($m \neq 0$)

1. $a = 0.04$, $m = 1$; $0 < \theta < 8\pi$ ("Archimedes' spiral")
2. $a = 0.20$, $m = 2$; $0 < \theta < 8\pi$ ("Fermat's spiral")
3. $a = 1.00$, $m = -1$; $1.00 < \theta < 6\pi$ ("hyperbolic spiral")
4. $a = 0.50$, $m = -2$; $0.25 < \theta < 6\pi$ ("lituus")

3.16.3. $r = a \cdot \cos(m\,\theta)$ $(x^2 + y^2)^{1/2} - a \cdot \cos[m \cdot \arctan(y/x)] = 0$

"Rhodonea" (also called "rose")

1. $a = 1.0$, $m = 4.0$; $0 < \theta < 2\pi$
2. $a = 1.0$, $m = 3.0$; $0 < \theta < \pi$

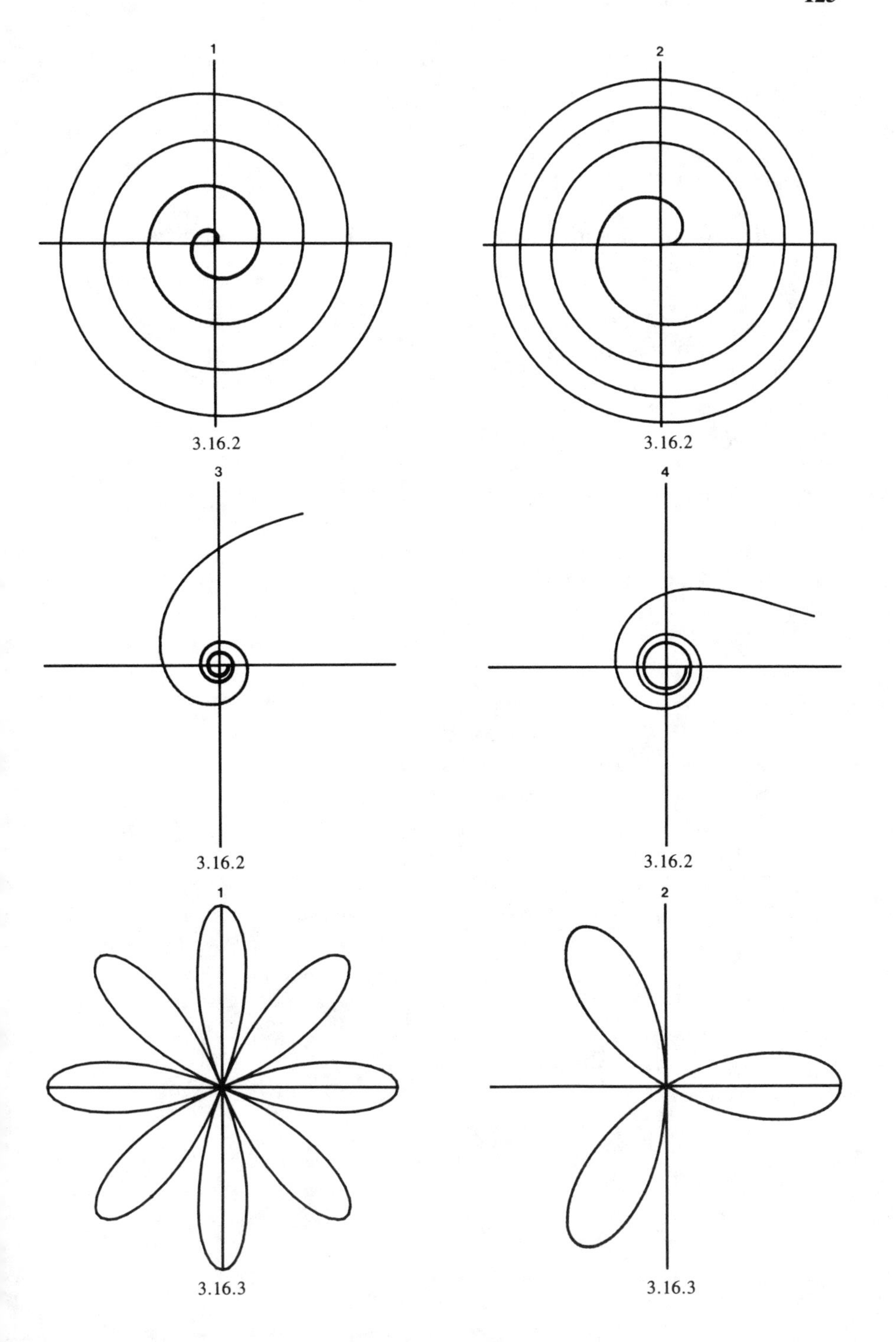

3.16.2

3.16.2

3.16.2

3.16.2

3.16.3

3.16.3

3.16.4. $r = a/\cos(m\,\theta)$ $(x^2 + y^2)^{1/2} - a/\cos[m \cdot \arctan(y/x)] = 0$
"Epi-spiral"

1. $a = 0.1$, $m = 4.0$; $0 < \theta < 2\pi$
2. $a = 0.1$, $m = 3.0$; $0 < \theta < \pi$

3.16.5. $r = (4b\,\theta)^{1/2} + a$ $[(x^2 + y^2)^{1/2} - a]^2 - 4b \cdot \arctan(y/x) = 0$
"Parabolic spiral"

1. $a = 0.1$, $b = 0.01$; $0 < \theta < 6\pi$
2. $a = -0.1$, $b = 0.01$; $0 < \theta < 6\pi$

3.16.6. $r = a \cdot \sin\theta/\theta$ $(x^2 + y^2)\arctan(y/x) - ay = 0$
"Cochleoid"

1. $a = 1.0$; $-6\pi < \theta < 6\pi$

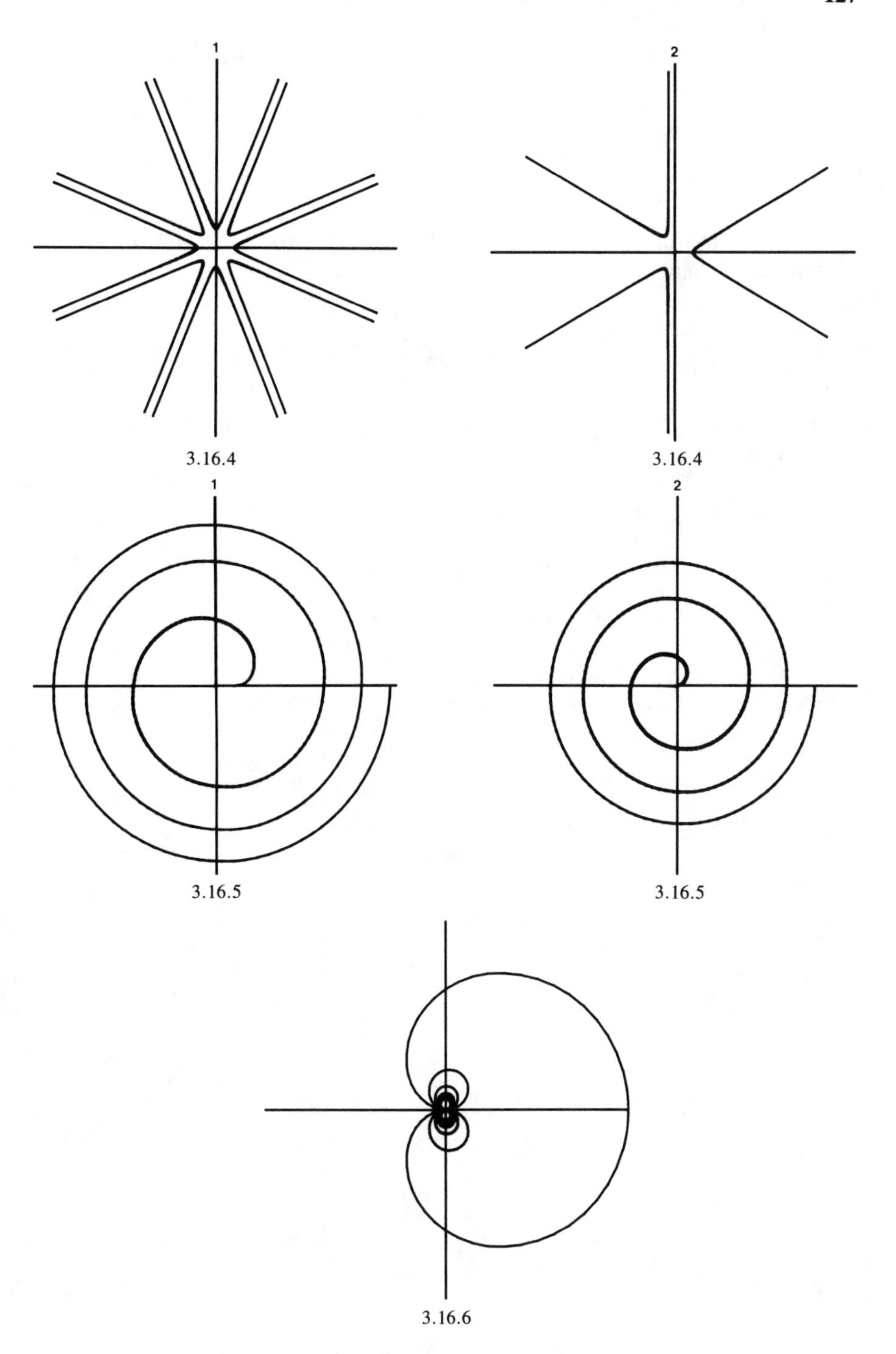

3.16.4

3.16.4

3.16.5

3.16.5

3.16.6

3.16.7. $r = a/\sinh(b\,\theta)$ $(x^2 + y^2)^{1/2} - a/\cosh[b \cdot \arctan(y/x)] = 0$
"Spiral of Poinsot"

1. $a = 0.5,\ b = 1.0;\ -2\pi < \theta < 2\pi$
2. $a = 0.5,\ b = 0.5;\ -2\pi < \theta < 2\pi$

3.16.8. $r = a/\cosh(b\,\theta)$ $(x^2 + y^2)^{1/2} - a/\sinh[b \cdot \arctan(y/x)] = 0$
"Spiral of Poinsot"

1. $a = 1.0,\ b = 1.0,\ -2\pi < \theta < 2\pi$
2. $a = 1.0,\ b = 0.5,\ -2\pi < \theta < 2\pi$

3.16.9. $r = a(1 + \theta^2)^{1/2}$ $(x^2 + y^2)^{1/2} - a\{1 + [\arctan(y/x)]^2\}^{1/2} = 0$
"Involute of a circle"

1. $a = 0.1;\ 0 < \theta < 2\pi$
2. $a = 0.2;\ 0 < \theta < 3\pi/2$

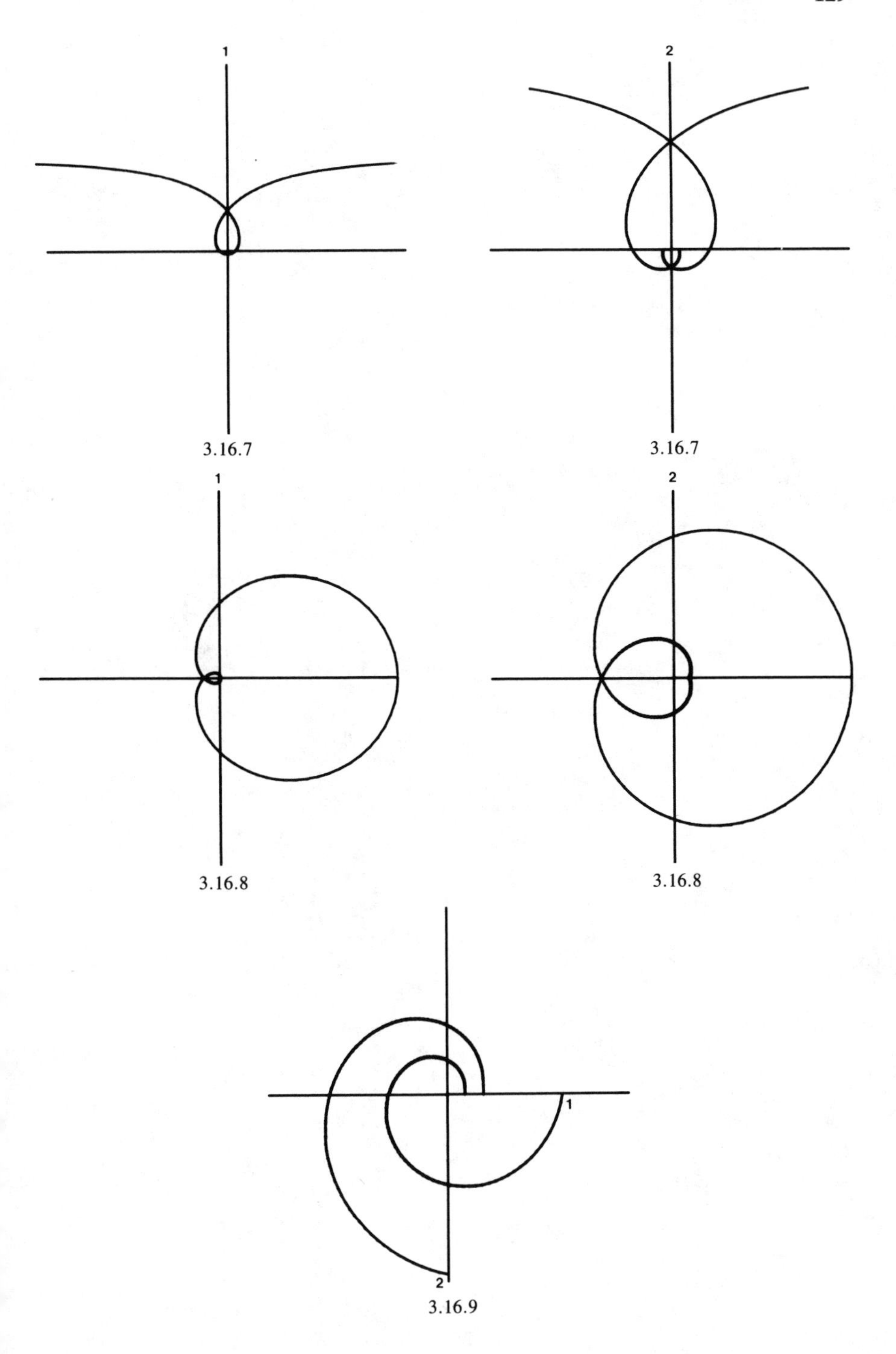

3.16.7

3.16.7

3.16.8

3.16.8

3.16.9

Chapter 4

POLYNOMIAL SETS

The polynomial sets illustrated in this chapter are treated in detail in Abramowitz and Stegun[1] or Beyer.[2] Because efficient calculation of the curves is achieved by using the recurrence relations given in these references, the relations are repeated here for anyone who may wish to generate the curves for their own purposes.

4.1. ORTHOGONAL POLYNOMIALS

4.1.1. Legendre Polynomials $P_n(x)$

Domain: $-1 < x < 1$

Recurrence relation: $P_{n+1}(x) = [(2n + 1)\, x\, P_n(x) - nP_{n-1}(x)]/(n + 1)$ with

$P_0(x) = 1$

$P_1(x) = x$

0. $P_0(x)$
1. $P_1(x)$
2. $P_2(x)$
3. $P_3(x)$
4. $P_4(x)$
5. $P_5(x)$
6. $P_6(x)$
7. $P_7(x)$
8. $P_8(x)$
9. $P_9(x)$

4.1.2. Chebyshev Polynomials of the First Kind $T_n(x)$

Domain: $-1 < x < 1$

Recurrence relation: $T_{n+1}(x) = 2xT_n(x) - T_{n-1}(x)$ with

$T_0(x) = 1$

$T_1(x) = x$

0. $T_0(x)$
1. $T_1(x)$
2. $T_2(x)$
3. $T_3(x)$
4. $T_4(x)$
5. $T_5(x)$
6. $T_6(x)$
7. $T_7(x)$

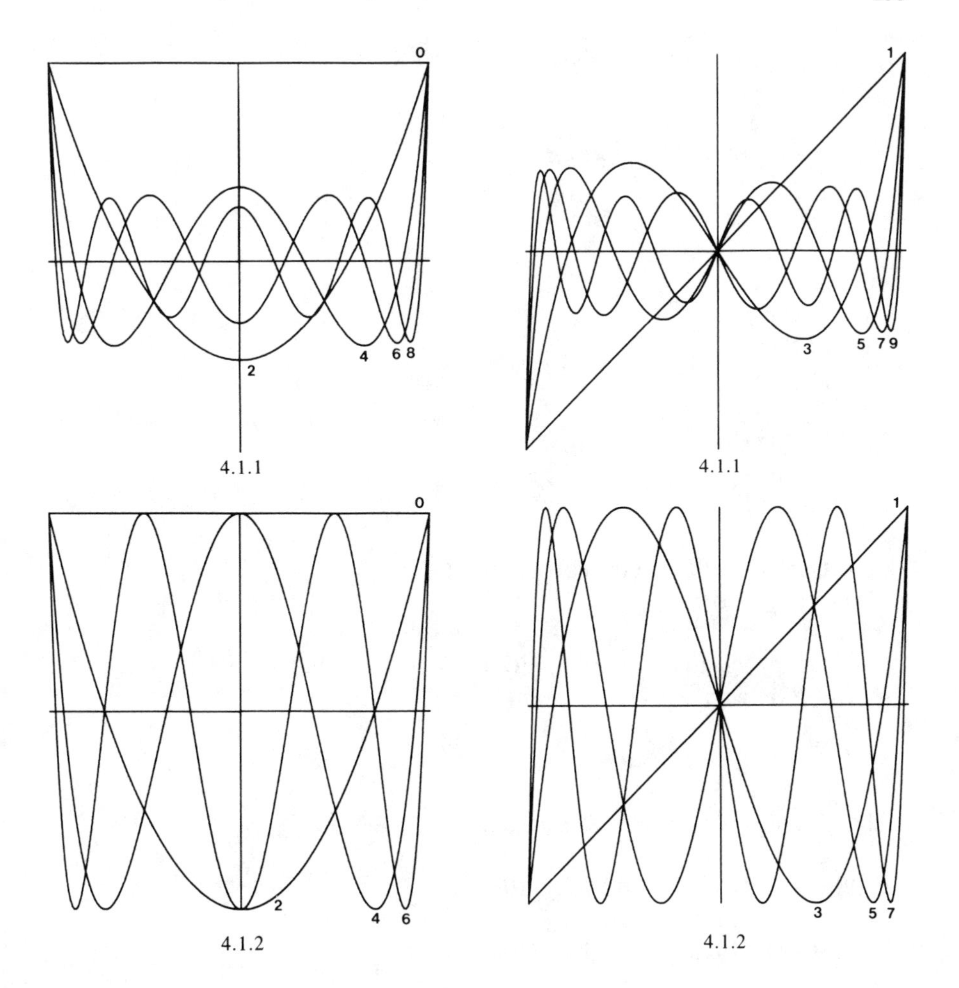

4.1.1

4.1.1

4.1.2

4.1.2

4.1.3. Chebyshev Polynomials of the Second Kind $U_n(x)$

Domain: $-1 < x < 1$

Recurrence relation: $U_{n+1}(x) = 2xU_n(x) - U_{n-1}(x)$ with
$U_0(x) = 1$
$U_1(x) = 2x$

0. $0.2\ U_0(x)$
1. $0.2\ U_1(x)$
2. $0.2\ U_2(x)$
3. $0.2\ U_3(x)$
4. $0.2\ U_4(x)$
5. $0.2\ U_5(x)$
6. $0.2\ U_6(x)$
7. $0.2\ U_7(x)$

4.1.4. Generalized Laguerre Polynomials $L_n^a(x)$

Domain: $x > 0$

Recurrence relation: $L^a_{n+1}(x) = [(2n + a + 1 - x)L_n^a(x) - (n + a)L^a_{n-1}(x)]/(n + 1)$ with
$L_0^a(x) = 1$
$L_1^a(x) = 1 - x + a$

1-0. $0.1\ L_0^1(10x)$ 2-0. $0.1\ L_0^2(10x)$
1-1. $0.1\ L_1^1(10x)$ 2-1. $0.1\ L_1^2(10x)$
1-2. $0.1\ L_2^1(10x)$ 2-2. $0.1\ L_2^2(10x)$
1-3. $0.1\ L_3^1(10x)$ 2-3. $0.1\ L_3^2(10x)$
1-4. $0.1\ L_4^1(10x)$ 2-4. $0.1\ L_4^2(10x)$

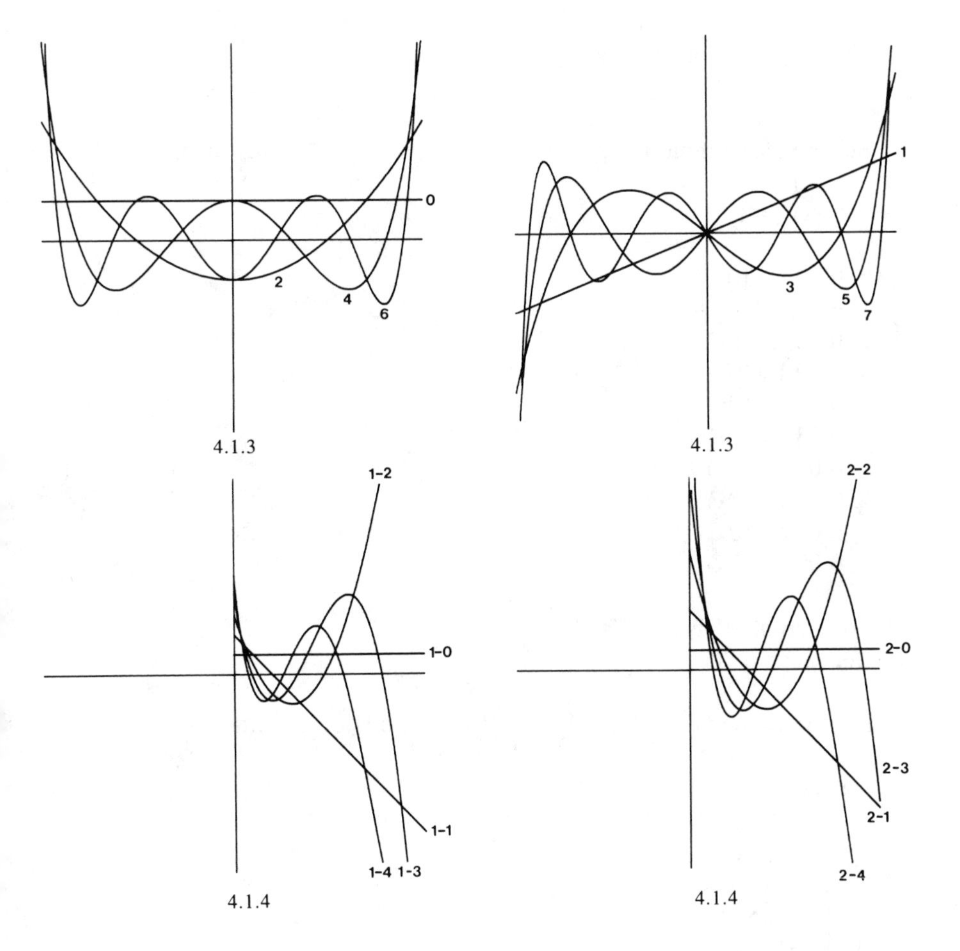

4.1.3

4.1.3

4.1.4

4.1.4

4.1.5. Laguerre Polynomials $L_n(x)$
Domain: $x > 0$

Recurrence relation: $L_{n+1}(x) = [(2n + 1 - x)L_n(x) - nL_{n-1}(x)]/(n + 1)$ with
$L_0(x) = 1$
$L_1(x) = 1 - x$

0. $0.10\ L_0(10x)$
1. $0.10\ L_1(10x)$
2. $0.10\ L_2(10x)$
3. $0.10\ L_3(10x)$
4. $0.10\ L_4(10x)$
5. $0.05\ L_5(10x)$
6. $0.05\ L_6(10x)$
7. $0.05\ L_7(10x)$
8. $0.05\ L_8(10x)$
9. $0.05\ L_9(10x)$

4.1.6. Hermite Polynomials $H_n(x)$
Domain: $x > 0$

Recurrence relation: $H_{n+1}(x) = 2xH_n(x) - 2nH_{n-1}(x)$ with
$H_0(x) = 1$
$H_1(x) = 2x$

0. $0.1\ H_0(5x)$
1. $0.1\ H_1(5x)$
2. $0.1\ H_2(5x)/2^3$
3. $0.1\ H_3(5x)/3^3$
4. $0.1\ H_4(5x)/4^3$
5. $0.1\ H_5(5x)/5^3$

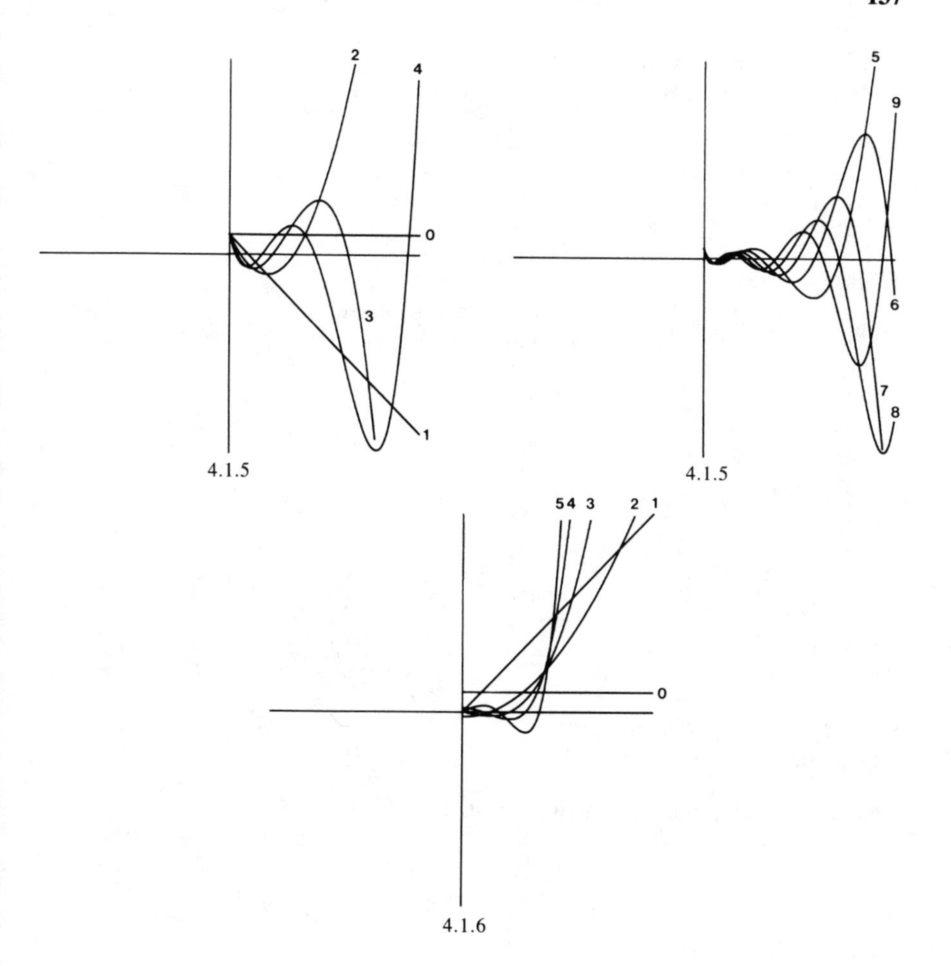

4.1.5

4.1.5

4.1.6

4.1.7. Gegenbauer Polynomials $C_n^a(x)$

Domain: $-1 < x < 1$

Recurrence relation: $C^a_{n+1}(x) = [2(n + a)xC_n^a - (n + 2a - 1)C^a_{n-1}]/(n + 1)$ with

$C_0^a(x) = 1$

$C_1^a(x) = 2ax$

Special cases

a = 1.0 gives Chebyshev polynomials of the second kind

a = 0.5 gives Legendre polynomials

0. $C_0^{1/4}(x)$
1. $C_1^{1/4}(x)$
2. $C_2^{1/4}(x)$
3. $C_3^{1/4}(x)$
4. $C_4^{1/4}(x)$
5. $C_5^{1/4}(x)$
6. $C_6^{1/4}(x)$
7. $C_7^{1/4}(x)$
8. $C_8^{1/4}(x)$
9. $C_9^{1/4}(x)$

4.1.8. Jacobi Polynomials $P_n^{a,b}(x)$

Domain: $-1 < x < 1$

Recurrence relation:

$$P^{a,b}_{n+1} = \{(2n + a + b + 1)[(a^2 - b^2) + (2n + a + b + 2)(2n + a + b)x]P^{a,b}_n - 2(n + a)(n + b)(2n + a + b + 2)P^{a,b}_{n-1}\}/[2(n + 1)(n + a + b + 1)(2n + a + b)]$$

with

$P_0^{a,b} = 1$

$P_1^{a,b} = [a - b + (a + b + 2)x]/2$

1-0. $P_0^{-1/2,1/2}(x)$
1-1. $P_1^{-1/2,1/2}(x)$
1-2. $P_2^{-1/2,1/2}(x)$
1-3. $P_3^{-1/2,1/2}(x)$
1-4. $P_4^{-1/2,1/2}(x)$
1-5. $P_5^{-1/2,1/2}(x)$

2-0. $P_0^{-1/2,1}(x)$
2-1. $P_1^{-1/2,1}(x)$
2-2. $P_2^{-1/2,1}(x)$
2-3. $P_3^{-1/2,1}(x)$
2-4. $P_4^{-1/2,1}(x)$
2-5. $P_5^{-1/2,1}(x)$

3-0. $P_0^{1,-1/2}(x)$
3-1. $P_1^{1,-1/2}(x)$
3-2. $P_2^{1,-1/2}(x)$
3-3. $P_3^{1,-1/2}(x)$
3-4. $P_4^{1,-1/2}(x)$
3-5. $P_5^{1,-1/2}(x)$

4-0. $P_0^{1,1/2}(x)$
4-1. $P_1^{1,1/2}(x)$
4-2. $P_2^{1,1/2}(x)$
4-3. $P_3^{1,1/2}(x)$
4-4. $P_4^{1,1/2}(x)$
4-5. $P_5^{1,1/2}(x)$

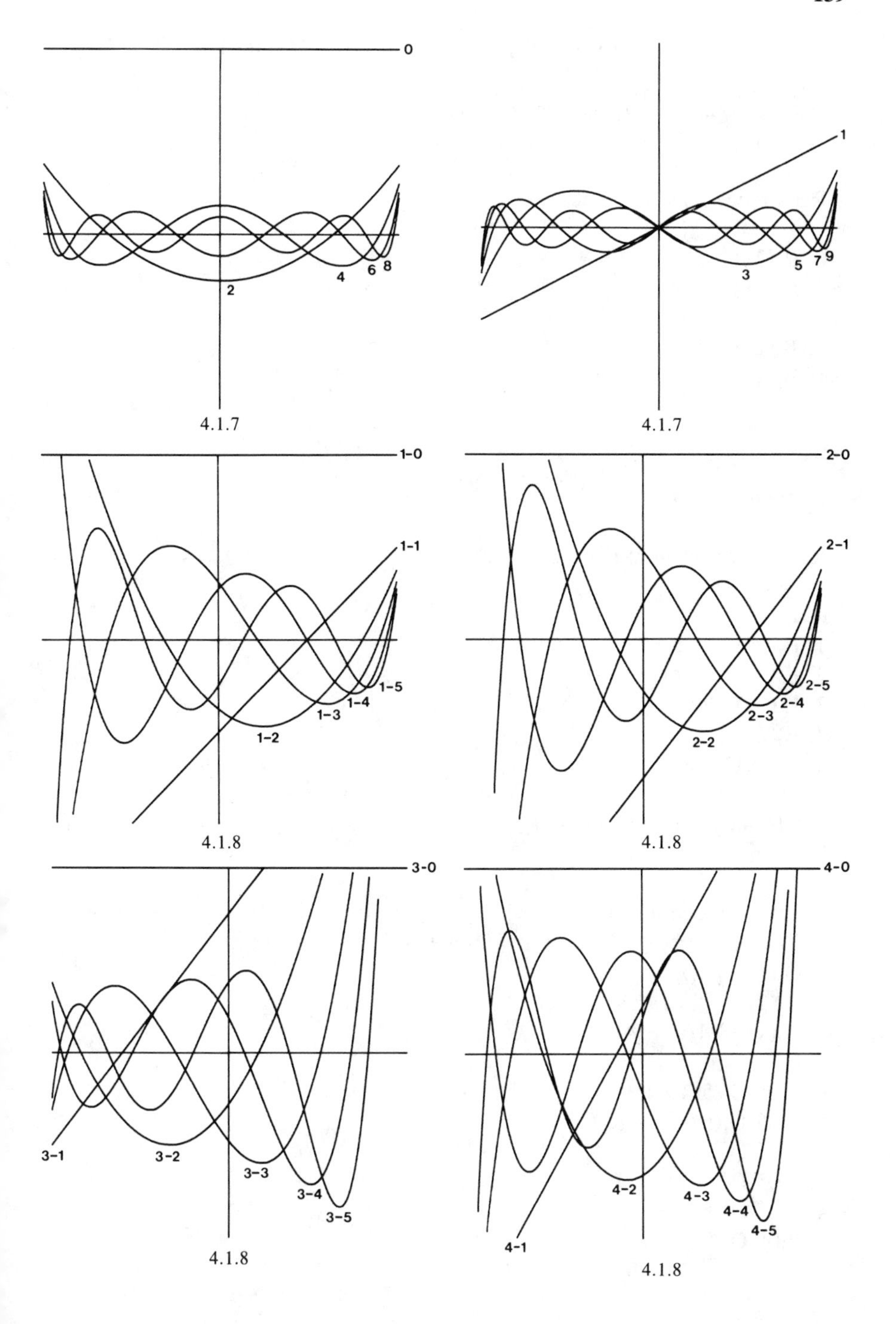

4.1.7

4.1.7

4.1.8

4.1.8

4.1.8

4.1.8

4.2. NONORTHOGONAL POLYNOMIALS

4.2.1. Bernoulli Polynomials $B_n(x)$

Domain: $x > 0$

Recurrence relation: none

0. $B_0(2x)$
1. $B_1(2x)$
2. $B_2(2x)$
3. $B_3(2x)$
4. $B_4(2x)$
5. $B_5(2x)$

4.2.2. Euler Polynomials $E_n(x)$

Domain: $x > 0$

Recurrence relation: none

0. $E_0(2x)$
1. $E_1(2x)$
2. $E_2(2x)$
3. $E_3(2x)$
4. $E_4(2x)$
5. $E_5(2x)$

4.2.3. Neumann Polynomials $O_n(x)$

Domain: $x > 0$

Recurrence relation (for $n > 1$): $O_{n+1}(x) = (n + 1)(2/x)O_n(x) - [(n + 1)/(n-1)]O_{n-1}(x) + (2n/x)\sin^2(n\pi/2)$ with

$O_0(x) = 1/x$

$O_1(x) = 1/x^2$

$O_2(x) = 1/x + 4/x^3$

0. $0.05\ O_0(5x)$
1. $0.05\ O_1(5x)$
2. $0.05\ O_2(5x)$
3. $0.05\ O_3(5x)$
4. $0.05\ O_4(5x)$
5. $0.05\ O_5(5x)$

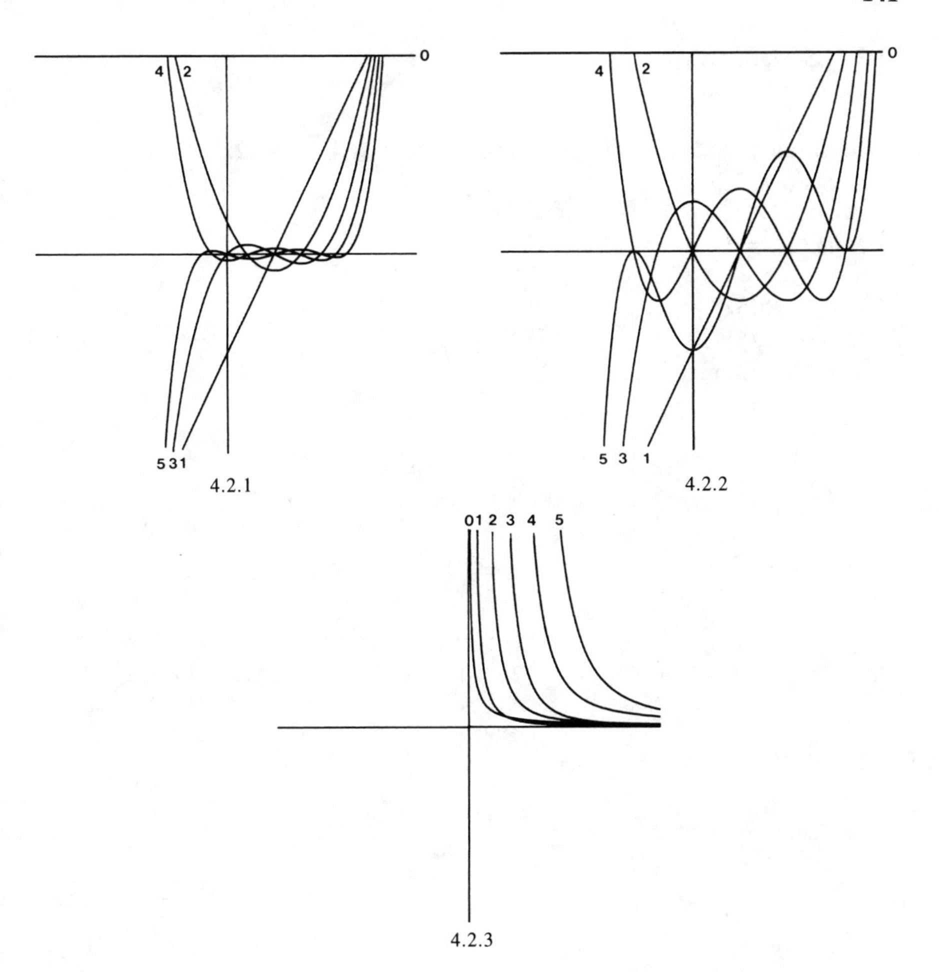

4.2.1

4.2.2

4.2.3

4.2.4. Schlafli Polynomials $S_n(x)$

Domain: $x > 0$

Recurrence relation from Neumann polynomials: $S_n(x) = [2xO_n(x) - 2\cos^2(n\pi/2)]/n$

Note: $S_0 = 0$

1. $0.05\ S_1(5x)$
2. $0.05\ S_2(5x)$
3. $0.05\ S_3(5x)$
4. $0.05\ S_4(5x)$
5. $0.05\ S_5(5x)$

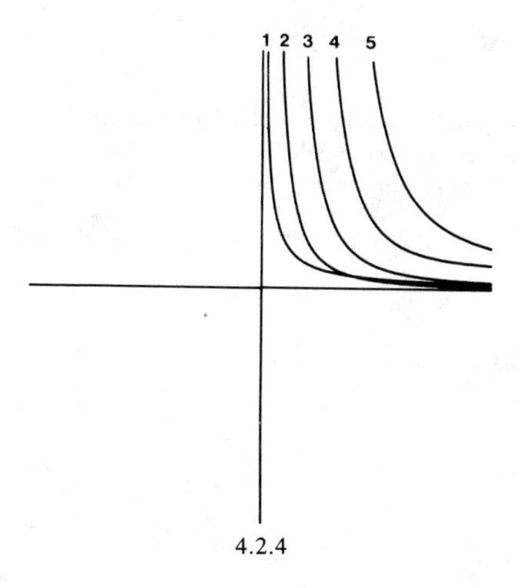

4.2.4

REFERENCES

1. **Abramowitz, M. and Stegun, I. A., Eds.,** Handbook of Mathematical Functions, National Bureau of Standards, U.S. Department of Commerce, Washington, D.C., 1964.
2. **Beyer, W. H., Ed.,** *Handbook of Mathematical Sciences,* 6th Ed., CRC Press, Boca Raton, Florida, 1987.

Chapter 5

SPECIAL FUNCTIONS IN MATHEMATICAL PHYSICS

The curves in this chapter are found in Abramowitz and Stegun,[1] and the names and notation used here conform with that reference. The approximations necessary to compute these curves are also given there; for purposes of illustrating the curves, the approximations were encoded into computer algorithms such that accuracy was attained to at least three significant figures for all plotted points of a curve. Such accuracy is sufficient for illustrative purposes and was efficiently achieved in all cases. The curves shown in this chapter are only representative, and the interested reader should, when necessary, consult the above reference, or similar ones such as Jahnke and Emde,[2] Beyer,[3] and Gradshteyn and Ryzhik,[4] for a complete treatment of these curves. The reader should be aware that many of the functions are defined for a complex argument while they are only plotted for a real argument in this chapter, thus showing only a vertical slice of the three-dimensional surface over the complex plane.

5.1. EXPONENTIAL AND RELATED INTEGRALS

5.1.1. Exponential Integral $E_n(x) = \int_1^\infty (e^{-xt}/t^n)dt$

Domain: $x > 0$

Recurrence relation: $E_{n+1}(x) = (1/n)[e^{-x} - xE_n(x)]$ $n = 1,2,3,\ldots$

0. $E_0(x)$ ($E_0(x) = e^{-x}/x$)
1. $E_1(x)$
2. $E_2(x)$
3. $E_3(x)$
4. $E_4(x)$
5. $E_5(x)$
6. $E_6(x)$
7. $E_7(x)$

5.1.2. Exponential Integral $Ei(x) = -\int_{-x}^\infty (e^{-t}/t)dt$

Domain: $x > 0$

1. 0.5 $Ei(x)$

5.1.3. Alpha Integral $\alpha_n(x) = \int_1^\infty t^n e^{-xt}dt$

Domain: $x > 0$

Recurrence relation: $\alpha_{n+1}(x) = (1/x)[e^{-x} + (n + 1)\,\alpha_n(x)]$ $n = 0,1,2,\ldots$

0. 0.2 $\alpha_0(5x)$(α_0 $(x) = e^{-x}/x$)
1. 0.2 $\alpha_1(5x)$
2. 0.2 $\alpha_2(5x)$
3. 0.2 $\alpha_3(5x)$
4. 0.2 $\alpha_4(5x)$
5. 0.2 $\alpha_5(5x)$
6. 0.2 $\alpha_6(5x)$
7. 0.2 $\alpha_7(5x)$

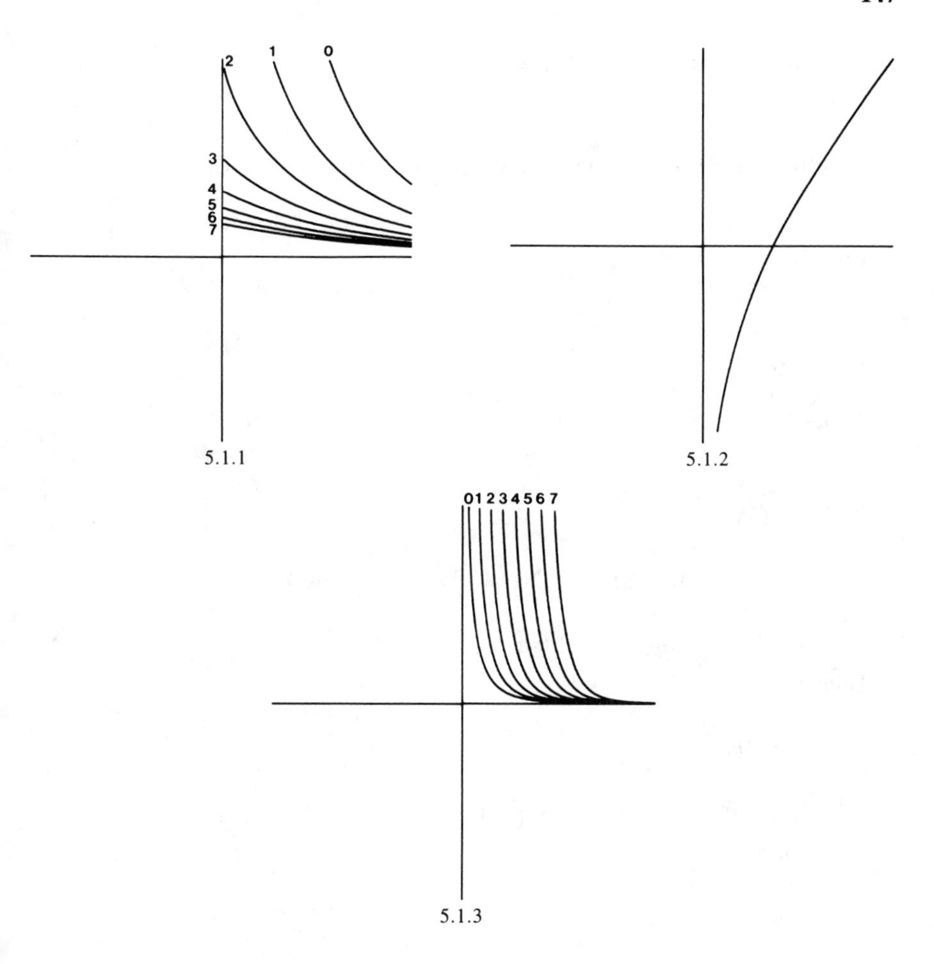

5.1.1

5.1.2

5.1.3

5.1.4. Beta Integral $\beta_n(x) = \int_{-1}^{1} t^n e^{-xt} dt$

Domain: $x > 0$

Recurrence relation: $\beta_{n+1}(x) = (1/x)[(-1)^{n+1}e^x - e^{-x} + (n+1)\beta_n(x)]$
$n = 0,1,2,\ldots$

with $\beta_0(x) = (2/x)\sinh(x)$

0. $0.1\ \beta_0(5x)$
1. $0.1\ \beta_1(5x)$
2. $0.1\ \beta_2(5x)$
3. $0.1\ \beta_3(5x)$
4. $0.1\ \beta_4(5x)$
5. $0.1\ \beta_5(5x)$
6. $0.1\ \beta_6(5x)$
7. $0.1\ \beta_7(5x)$

5.2. SINE AND COSINE INTEGRALS

5.2.1. Sine Integral $\mathrm{Si}(x) = \int_0^x (\sin t/t)dt$

Domain: $x > 0$

1. $0.5\ \mathrm{Si}(20x)$

5.2.2. Cosine Integral $\mathrm{Ci}(x) = \gamma + \ln(x) + \int_0^x [(\cos t - 1)/t]dt$

Domain: $x > 0$

1. $\mathrm{Ci}(20x)$

5.3. GAMMA AND RELATED FUNCTIONS

5.3.1. Gamma Function $\Gamma(x) = \int_0^\infty t^{x-1}e^{-t}dt$

Also called "Euler's Integral of the Second Kind"

Domain: $-\infty < x < \infty$

Recurrence relation: $\Gamma(x + 1) = x\Gamma(x)$

1. $0.2\ \Gamma(5x)$

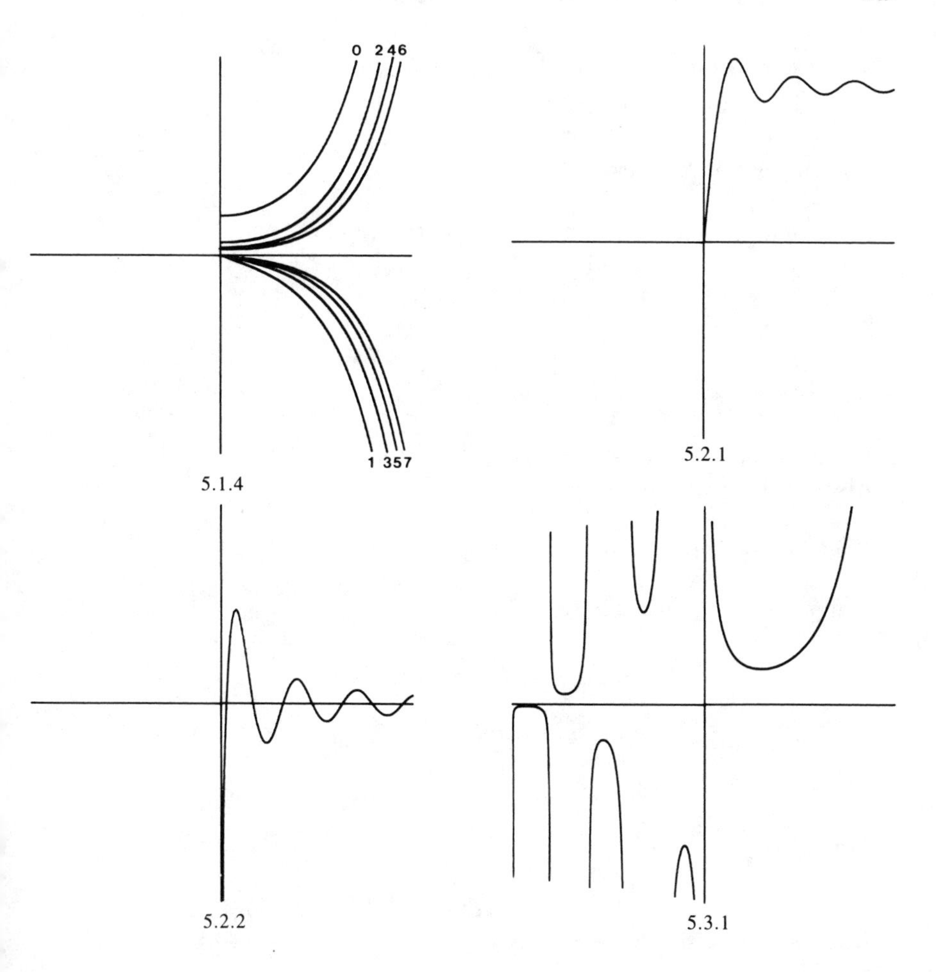

5.1.4

5.2.1

5.2.2

5.3.1

5.3.2. Beta Function $B(x,w) = \int_0^1 t^{x-1}(1-t)^{w-1}dt$

Also called "Euler's Integral of the First Kind"

Domain: $-\infty < x < \infty$

Relation to Gamma Function: $B(x,w) = \Gamma(x)\Gamma(w)/\Gamma(x + w)$

1. 0.5 B(5x,1)
2. 0.5 B(5x,2)
3. 0.5 B(5x,3)
4. 0.5 B(5x,4)

5.3.3. Psi Function $\psi(x) = [d\Gamma(x)/dx]/\Gamma(x)$

Also called "Digamma Function"

Domain: $-\infty < x < \infty$

1. 0.2 ψ(5x)

5.4. ERROR FUNCTIONS

5.4.1. Error Function $\mathrm{Erf}(x) = (2/\pi^{1/2})\int_0^x \exp(-t^2)dt$

Domain: $-\infty < x < \infty$

1. Erf(2x)

5.4.2. Complementary Error Function $\mathrm{Erfc}(x) = 1 - \mathrm{Erf}(x)$

Domain: $-\infty < x < \infty$

1. 0.5 Erfc(2x)

5.5. FRESNEL INTEGRALS

5.5.1. First Fresnel Integral $S(x) = \int_0^x \sin(\pi t^2/2)dt$

Domain: $-\infty < x < \infty$

1. S(5x)

5.5.2. Second Fresnel Integral $C(x) = \int_0^x \cos(\pi t^2/2)dt$

Domain: $-\infty < x < \infty$

1. C(5x)

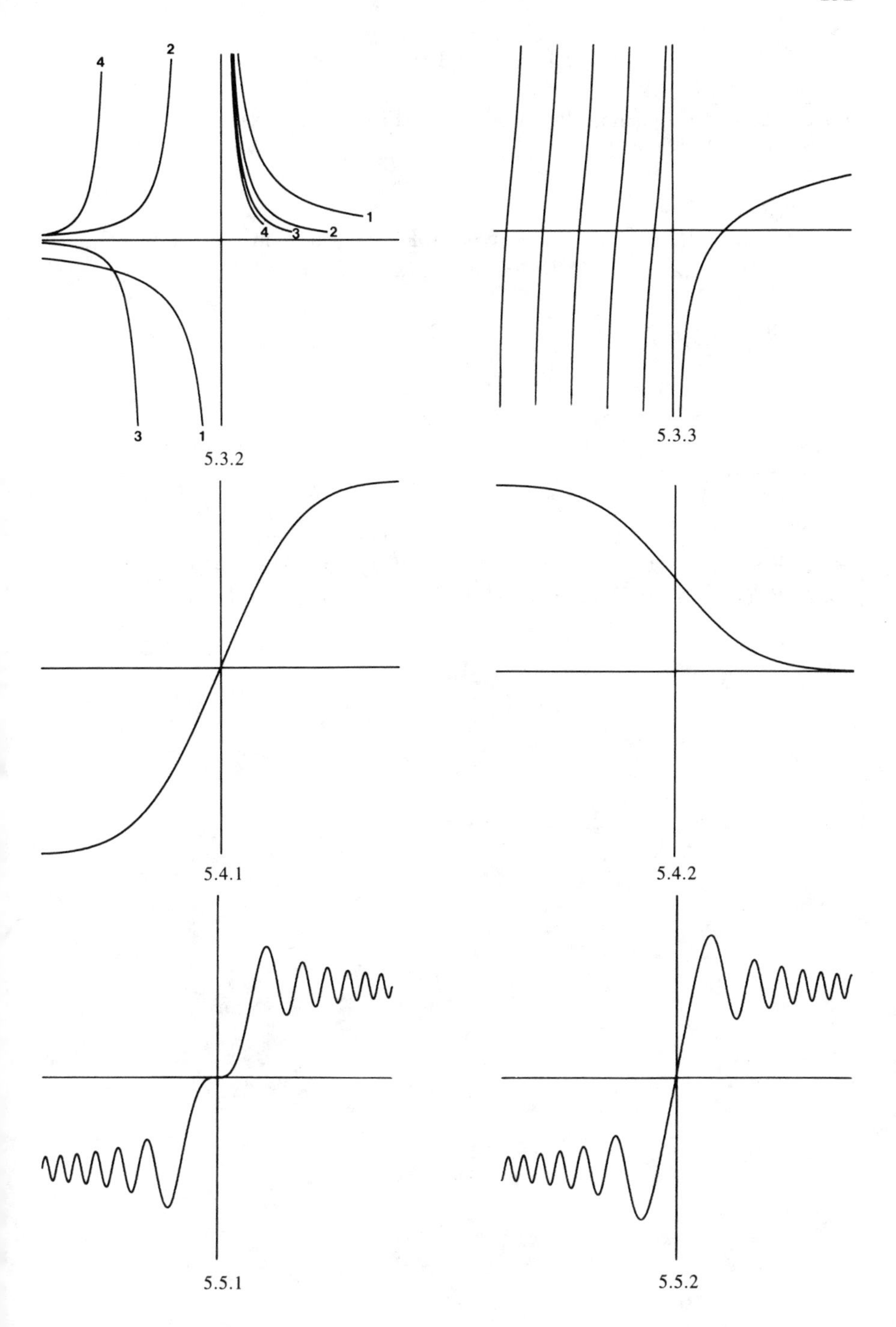

5.3.2

5.3.3

5.4.1

5.4.2

5.5.1

5.5.2

5.6. LEGENDRE FUNCTIONS

5.6.1. Associated Legendre Functions of the First Kind $P_n^m(x)$

Domain: $-1 < x < 1$

Recurrence relations:

$P_{n+1}^m(x) = [(2n + 1)\,xP_n^m - (n + m)P_{n-1}^m(x)]/(n - m + 1)$ $n = 1,2,3,\ldots$

$P_n^{m+1}(x) = (x^2 - 1)^{-1/2}[(n - m)\,xP_n^m(x) - (n + m)P_{n-1}^m(x)]$

$m = 0,1,2,\ldots$

with $P_0^0 = 1$

$P_1^0 = x$

Special case: P_n^0 = Legendre Polynomials

1-0. $P_0^0(x)$

1-1. $P_1^0(x)$ 2-1. $0.25\,P_1^1(x)$

1-2. $P_2^0(x)$ 2-2. $0.25\,P_2^1(x)$ 3-2. $0.20\,P_2^2(x)$

1-3. $P_3^0(x)$ 2-3. $0.25\,P_3^1(x)$ 3-3. $0.20\,P_3^2(x)$ 4-3. $0.10\,P_3^3(x)$

1-4. $P_4^0(x)$ 2-4. $0.25\,P_4^1(x)$ 3-4. $0.20\,P_4^2(x)$ 4-4. $0.10\,P_4^3(x)$

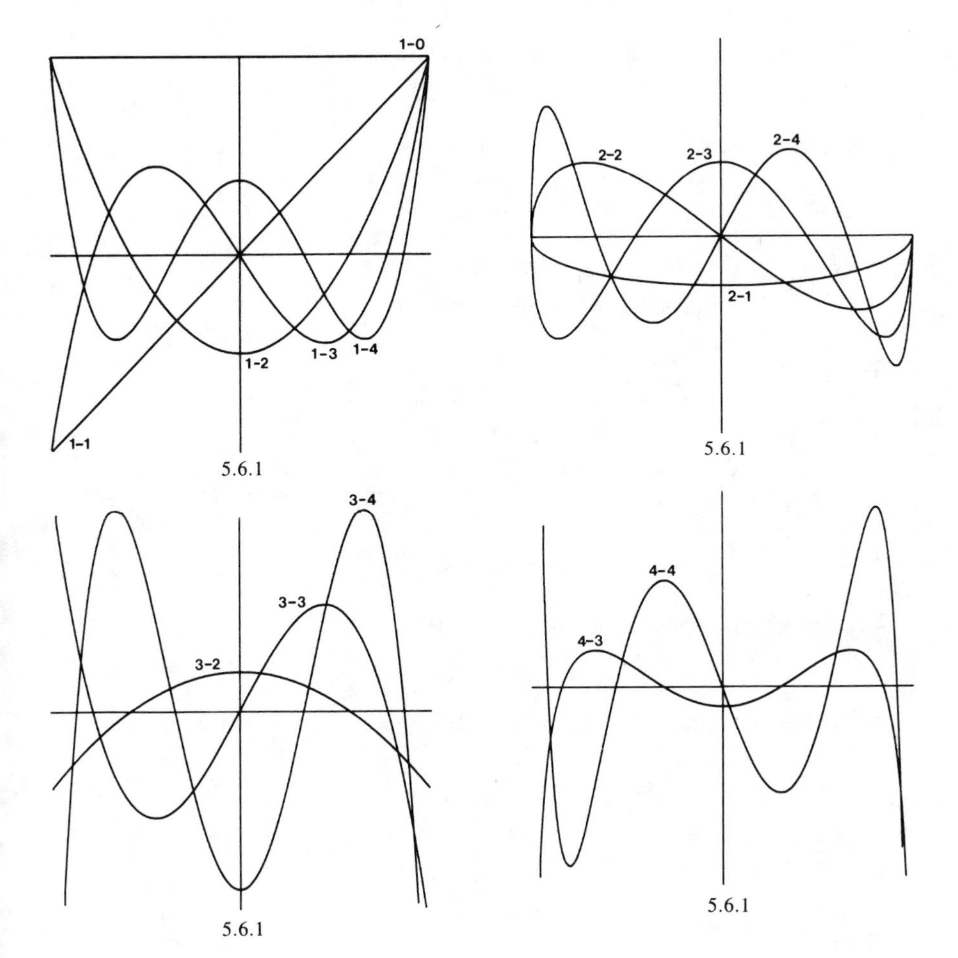

5.6.1

5.6.1

5.6.1

5.6.1

5.6.2. Associated Legendre Functions of the Second Kind $Q_n^m(x)$
Domain: $-1 < x < 1$

Recurrence relations:

$Q_{m+1}^m(x) = [(2n + 1)\, xQ_n^m - (n + m)\, Q_{n-1}^m(x)]/(n - m + 1)$ n = 1,2,3...

$Q_n^{m+1}(x) = (x^2 - 1)^{-1/2}[(n - m)\, xQ_n^m(x) - (n + m)\, Q_{n-1}^m(x)]$

m = 0,1,2,...

with $Q_0^0 = \ln[(1 + x)/(1 - x)]/2$

$Q_1^0 = (x/2)\ln[(1 + x)/(1 - x)] - 1$

1-0. $Q_0^0(x)$
1-1. $Q_1^0(x)$ 2-1. 0.25 $Q_1^1(x)$
1-2. $Q_2^0(x)$ 2-2. 0.25 $Q_2^1(x)$ 3-2 0.10 $Q_2^2(x)$
1-3. $Q_3^0(x)$ 2-3. 0.25 $Q_3^1(x)$ 3-3. 0.10 $Q_3^2(x)$ 4-3. 0.05 $Q_3^3(x)$
1-4. $Q_4^0(x)$ 2-4. 0.25 $Q_4^1(x)$ 3-4. 0.10 $Q_4^2(x)$ 4-4. 0.05 $Q_4^3(x)$

5.7. BESSEL FUNCTIONS

5.7.1. Bessel Functions of the First Kind $J_n(x)$
Domain: $x > 0$

Recurrence relation: $J_{n+1}(x) = (2n/x)J_n(x) - J_{n-1}(x)$ n = 0,1,2,...

Symmetry: $J_{-n}(x) = (-1)^n J_n(x)$

0. $J_0(20x)$
1. $J_1(20x)$
2. $J_2(20x)$
3. $J_3(20x)$
4. $J_4(20x)$
5. $J_5(20x)$

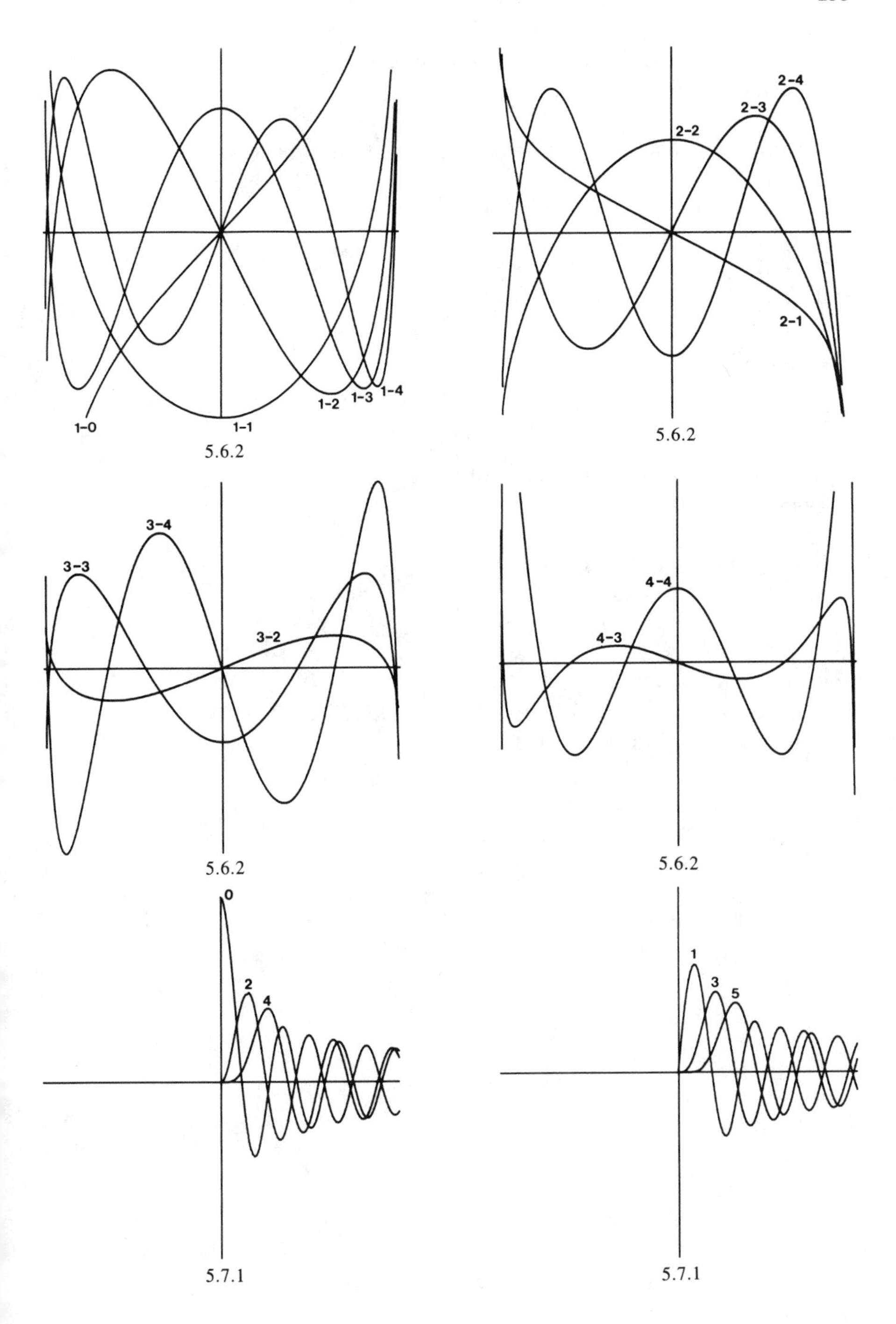

5.6.2

5.6.2

5.6.2

5.6.2

5.7.1

5.7.1

5.7.2. Bessel Functions of the Second Kind $Y_n(x)$

Domain: $x > 0$

Recurrence relation: $Y_{n+1}(x) = (2n/x)Y_n(x) - Y_{n-1}(x)$ $n = 0,1,2,\ldots$

Symmetry: $Y_{-n}(x) = (-1)^n Y_n(x)$

0. $Y_0(20x)$
1. $Y_1(20x)$
2. $Y_2(20x)$
3. $Y_3(20x)$
4. $Y_4(20x)$
5. $Y_5(20x)$

5.7.3. Hankel Functions $H_n^{(1)}(x)$ and $H_n^{(2)}(x)$

Domain: $x > 0$

Relation to Bessel Functions

$H_n^{(1)} = J_n(x) + iY_n(x)$

$H_n^{(2)} = J_n(x) - iY_n(x)$

Recurrence relation:

$H_{n+1}^{(1,2)}(x) = (2n/x)H_n^{(1,2)}(x) - H_{n-1}^{(1,2)}(x)$ $n = 0,1,2,\ldots$

Symmetry: $H_{-n}^{(1,2)}(x) = (-1)^n H_n^{(1,2)}(x)$

0. $|H_0^{(m)}(20x)|$ $m = 1,2$
1. $|H_1^{(m)}(20x)|$ $m = 1,2$
2. $|H_2^{(m)}(20x)|$ $m = 1,2$
3. $|H_3^{(m)}(20x)|$ $m = 1,2$
4. $|H_4^{(m)}(20x)|$ $m = 1,2$
5. $|H_5^{(m)}(20x)|$ $m = 1,2$

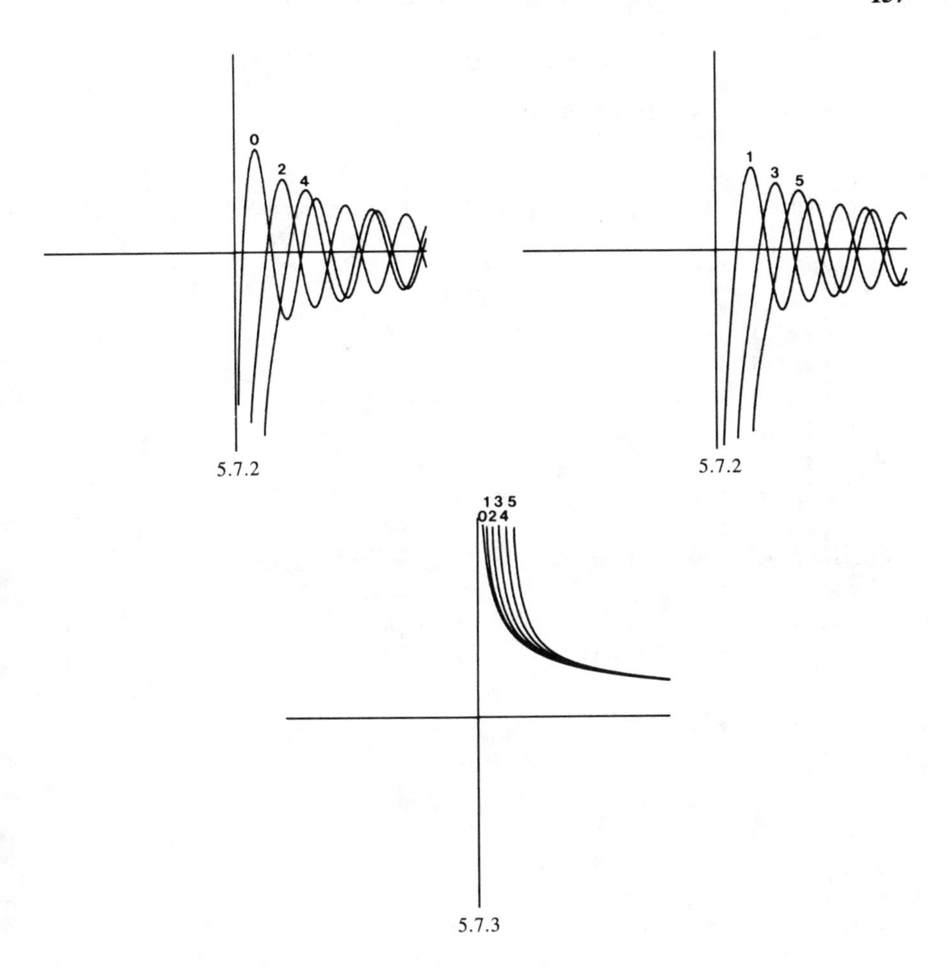

5.7.2

5.7.2

5.7.3

5.8. MODIFIED BESSEL FUNCTIONS

5.8.1. Modified Bessel Function $I_n(x)$

Domain: $x > 0$

Recurrence relation: $I_{n+1}(x) = I_{n-1}(x) - (2n/x)I_n(x)$ $n = 0,1,2,\ldots$

Symmetry: $I_{-n}(x) = I_n(x)$

0. $0.1\ I_0(5x)$
1. $0.1\ I_1(5x)$
2. $0.1\ I_2(5x)$
3. $0.1\ I_3(5x)$
4. $0.1\ I_4(5x)$
5. $0.1\ I_5(5x)$

5.8.2. Modified Bessel Function $K_n(x)$

Domain: $x > 0$

Recurrence relation: $K_{n+1}(x) = K_{n-1}(x) - (2n/x)K_n(x)$ $n = 0,1,2,\ldots$

Symmetry: $K_{-n}(x) = K_n(x)$

0. $0.1\ K_0(2x)$
1. $0.1\ K_1(2x)$
2. $0.1\ K_2(2x)$
3. $0.1\ K_3(2x)$
4. $0.1\ K_4(2x)$
5. $0.1\ K_5(2x)$

5.9. KELVIN FUNCTIONS

5.9.1. Kelvin Function $\mathrm{ber}_n(x)$

Domain: $x > 0$

Recurrence relation: $\mathrm{ber}_{n+1}(x) = -(2^{1/2}n/x)[\mathrm{ber}_n(x) - \mathrm{bei}_n(x)] - \mathrm{ber}_{n-1}(x)$
$n = 1,2,3,\ldots$

Symmetry: $\mathrm{ber}_{-n}(x) = (-1)^n\mathrm{ber}_n(x)$

0. $0.1\ \mathrm{ber}_0(8x)$
1. $0.1\ \mathrm{ber}_1(8x)$
2. $0.1\ \mathrm{ber}_2(8x)$
3. $0.1\ \mathrm{ber}_3(8x)$
4. $0.1\ \mathrm{ber}_4(8x)$
5. $0.1\ \mathrm{ber}_5(8x)$

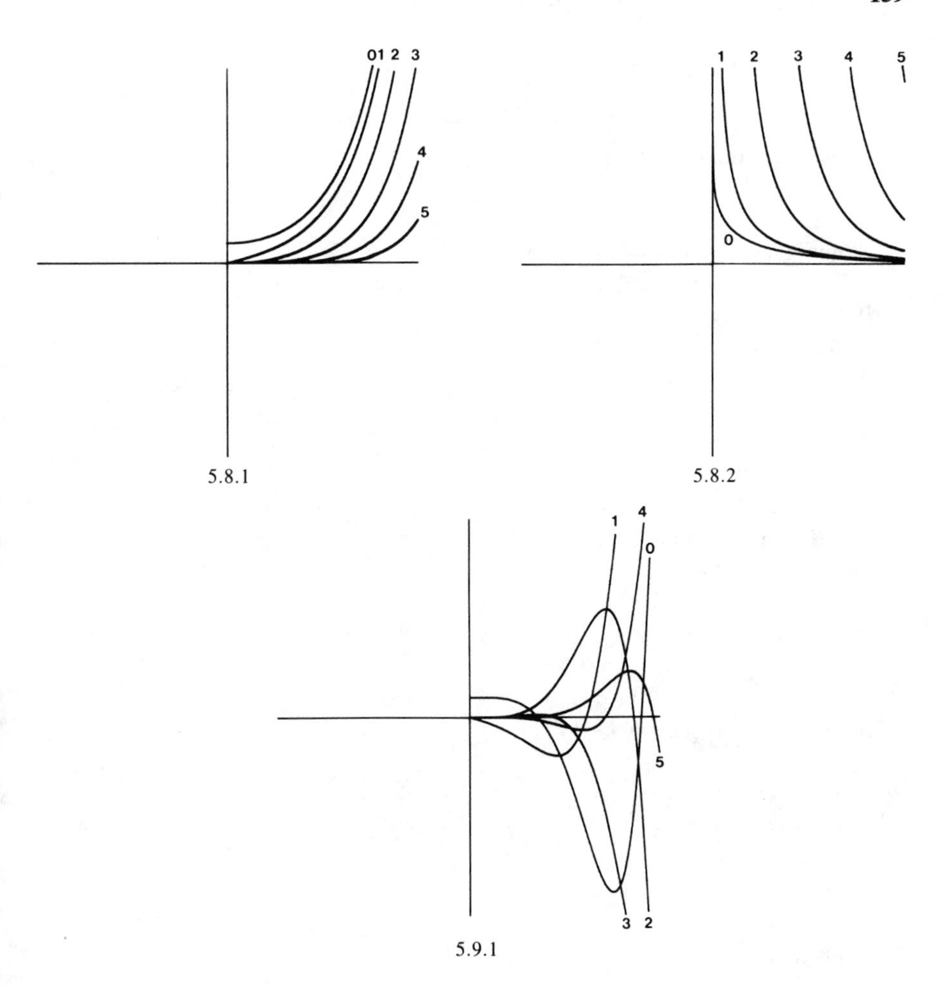

5.8.1

5.8.2

5.9.1

5.9.2 Kelvin Function $\mathrm{bei}_n(x)$
Domain: $x > 0$

Recurrence relation: $\mathrm{bei}_{n+1}(x) = -(2^{1/2}n/x)[\mathrm{bei}_n(x) + \mathrm{ber}_n(x)] - \mathrm{bei}_{n-1}(x)$
$n = 1,2,3,\ldots$

Symmetry: $\mathrm{bei}_{-n}(x) = (-1)^n \mathrm{bei}_n(x)$

0. $0.1\ \mathrm{bei}_0(8x)$
1. $0.1\ \mathrm{bei}_1(8x)$
2. $0.1\ \mathrm{bei}_2(8x)$
3. $0.1\ \mathrm{bei}_3(8x)$
4. $0.1\ \mathrm{bei}_4(8x)$
5. $0.1\ \mathrm{bei}_5(8x)$

5.9.3. Kelvin Function $\mathrm{ker}_n(x)$
Domain: $x > 0$

Recurrence relation: $\mathrm{ker}_{n+1}(x) = -(2^{1/2}n/x)[\mathrm{ker}_n(x) - \mathrm{kei}_n(x)] - \mathrm{ker}_{n-1}(x)$
$n = 1,2,3,\ldots$

Symmetry: $\mathrm{ker}_{-n}(x) = (-1)^n \mathrm{ker}_n(x)$

0. $\mathrm{ker}_0(8x)$
1. $\mathrm{ker}_1(8x)$
2. $\mathrm{ker}_2(8x)$
3. $\mathrm{ker}_3(8x)$
4. $\mathrm{ker}_4(8x)$
5. $\mathrm{ker}_5(8x)$

5.9.4. Kelvin Function $\mathrm{kei}_n(x)$
Domain: $x > 0$

Recurrence relation: $\mathrm{kei}_{n+1}(x) = -(2^{1/2}n/x)[\mathrm{kei}_n(x) + \mathrm{ker}_n(x)] - \mathrm{kei}_{n-1}(x)$
$n = 1,2,3,\ldots$

Symmetry: $\mathrm{kei}_{-n}(x) = (-1)^n \mathrm{kei}_n(x)$

0. $\mathrm{kei}_0(8x)$
1. $\mathrm{kei}_1(8x)$
2. $\mathrm{kei}_2(8x)$
3. $\mathrm{kei}_3(8x)$
4. $\mathrm{kei}_4(8x)$
5. $\mathrm{kei}_5(8x)$

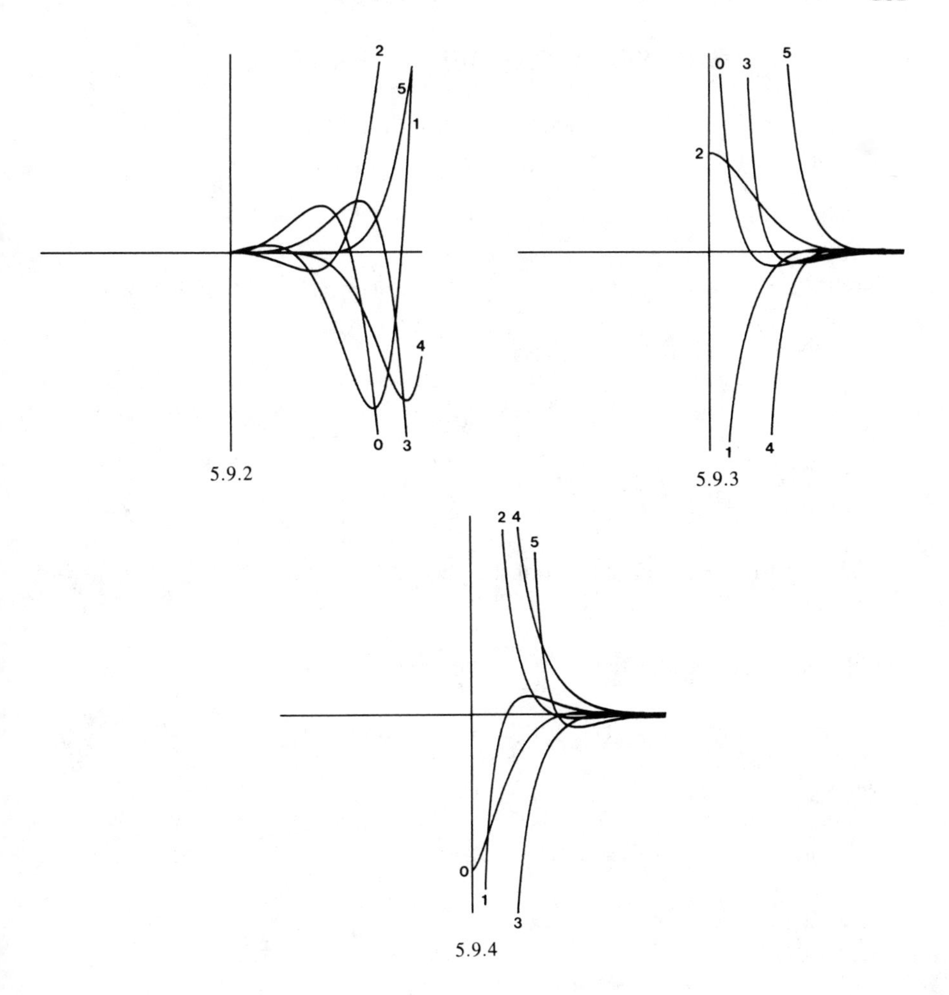

5.9.2

5.9.3

5.9.4

5.10. SPHERICAL BESSEL FUNCTIONS

5.10.1. Spherical Bessel Function of the First Kind $j_n(x)$

Domain: $x > 0$

Relation to Bessel Function: $j_n(x) = (\pi/2x)^{1/2}J_{n+1/2}(x)$

Recurrence relation: $j_{n+1}(x) = [(2n + 1)/x]j_n(x) - j_{n-1}(x)$ $n = 0,1,2,\ldots$

Symmetry: $j_{-n}(x) = (-1)^{-n}y_{n+1}(x)$

0. $j_0(20x)$
1. $j_1(20x)$
2. $j_2(20x)$
3. $j_3(20x)$

5.10.2. Spherical Bessel Function of the Second Kind $y_n(x)$

Domain: $x > 0$

Relation to Bessel Function: $y_n(x) = (\pi/2x)^{1/2}Y_{n+1/2}(x)$

Recurrence relation: $y_{n+1}(x) = [(2n + 1)/x]y_n(x) - y_{n-1}(x)$ $n = 0,1,2,\ldots$

Symmetry: $y_{-n}(x) = (-1)^{1-n}j_{n-1}(x)$

0. $y_0(20x)$
1. $y_1(20x)$
2. $y_2(20x)$
3. $y_3(20x)$

5.10.3. Spherical Bessel Function of the Third Kind $h_n^{(1)}(x)$ and $h_n^{(2)}(x)$

Domain: $x > 0$

Relation to Hankel Function: $h_n^{(1,2)}(x) = (\pi/2x)^{1/2}H_{n+1/2}^{(1,2)}(x)$

Recurrence relation: $h_{n+1}^{(1,2)}(x) = [(2n + 1)/x]h_n^{(1,2)}(x) - h_{n-1}^{(1,2)}(x)$
$n = 0,1,2,\ldots$

Symmetry: $h_{-n}^{(1)}(x) = i(-1)^{n+1}h_{n+1}^{(1)}(x)$; $h_{-n}^{(2)}(x) = i(-1)^n h_{n+1}^{(2)}(x)$

0. $|h_0^{(m)}(20x)|$ $m = 1,2$
1. $|h_1^{(m)}(20x)|$ $m = 1,2$
2. $|h_2^{(m)}(20x)|$ $m = 1,2$
3. $|h_3^{(m)}(20x)|$ $m = 1,2$

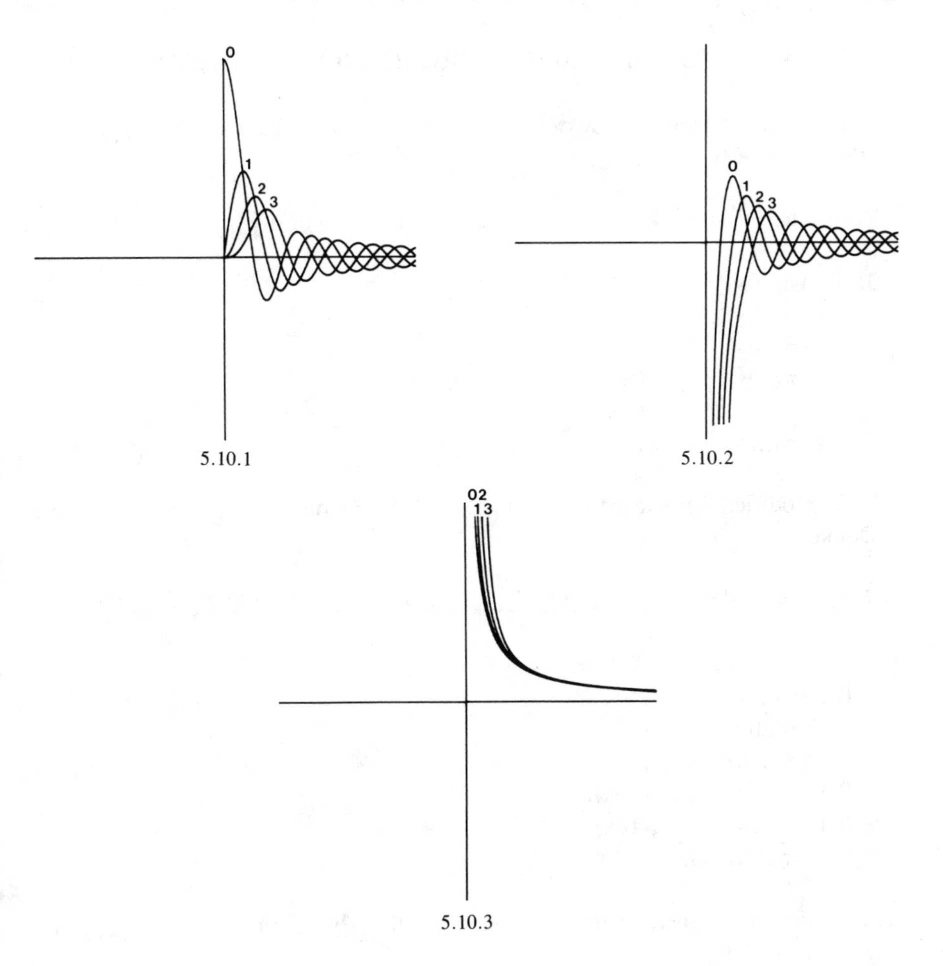

5.10.1

5.10.2

5.10.3

5.11. MODIFIED SPHERICAL BESSEL FUNCTIONS

5.11.1. Modified Spherical Bessel Function of the First Kind $(\pi/2x)^{1/2}I_{n+1/2}(x)$

Domain: $x > 0$

Recurrence relation: $I_{n+3/2}(x) = I_{n-1/2}(x) - [(2n + 1)/x)]I_{n+1/2}(x)$ $n = 0,1,2,\ldots$

0. $0.1\ (\pi/20x)^{1/2}\ I_{1/2}(10x)$
1. $0.1\ (\pi/20x)^{1/2}\ I_{3/2}(10x)$
2. $0.1\ (\pi/20x)^{1/2}\ I_{5/2}(10x)$
3. $0.1\ (\pi/20x)^{1/2}\ I_{7/2}(10x)$
4. $0.1\ (\pi/20x)^{1/2}\ I_{9/2}(10x)$
5. $0.1\ (\pi/20x)^{1/2}\ I_{11/2}(10x)$

5.11.2. Modified Spherical Bessel Function of the Second Kind $(\pi/2x)^{1/2}I_{-n-1/2}(x)$

Domain: $x > 0$

Recurrence relation: $I_{-n-3/2}(x) = I_{-n+1/2}(x) - [(2n + 1)/x)]I_{-n-1/2}(x)$
$n = 0,1,2,\ldots$

0. $0.1\ (\pi/20x)^{1/2}\ I_{-1/2}(10x)$
1. $0.1\ (\pi/20x)^{1/2}\ I_{-3/2}(10x)$
2. $0.1\ (\pi/20x)^{1/2}\ I_{-5/2}(10x)$
3. $0.1\ (\pi/20x)^{1/2}\ I_{-7/2}(10x)$
4. $0.1\ (\pi/20x)^{1/2}\ I_{-9/2}(10x)$
5. $0.1\ (\pi/20x)^{1/2}\ I_{-11/2}(10x)$

5.11.3. Modified Spherical Bessel Function of the Third Kind $(\pi/2x)^{1/2}K_{n+1/2}(x)$

Domain: $x > 0$

Recurrence relation: $K_{n+3/2}(x) = K_{n-1/2}(x) + [(2n + 1)/x)]K_{n+1/2}(x)$ $n = 0,1,2,\ldots$

Symmetry: $K_{-n-1/2}(x) = K_{n+1/2}(x)$

0. $(\pi/20x)^{1/2}\ K_{1/2}(10x)$
1. $(\pi/20x)^{1/2}\ K_{3/2}(10x)$
2. $(\pi/20x)^{1/2}\ K_{5/2}(10x)$
3. $(\pi/20x)^{1/2}\ K_{7/2}(10x)$
4. $(\pi/20x)^{1/2}\ K_{9/2}(10x)$
5. $(\pi/20x)^{1/2}\ K_{11/2}(10x)$

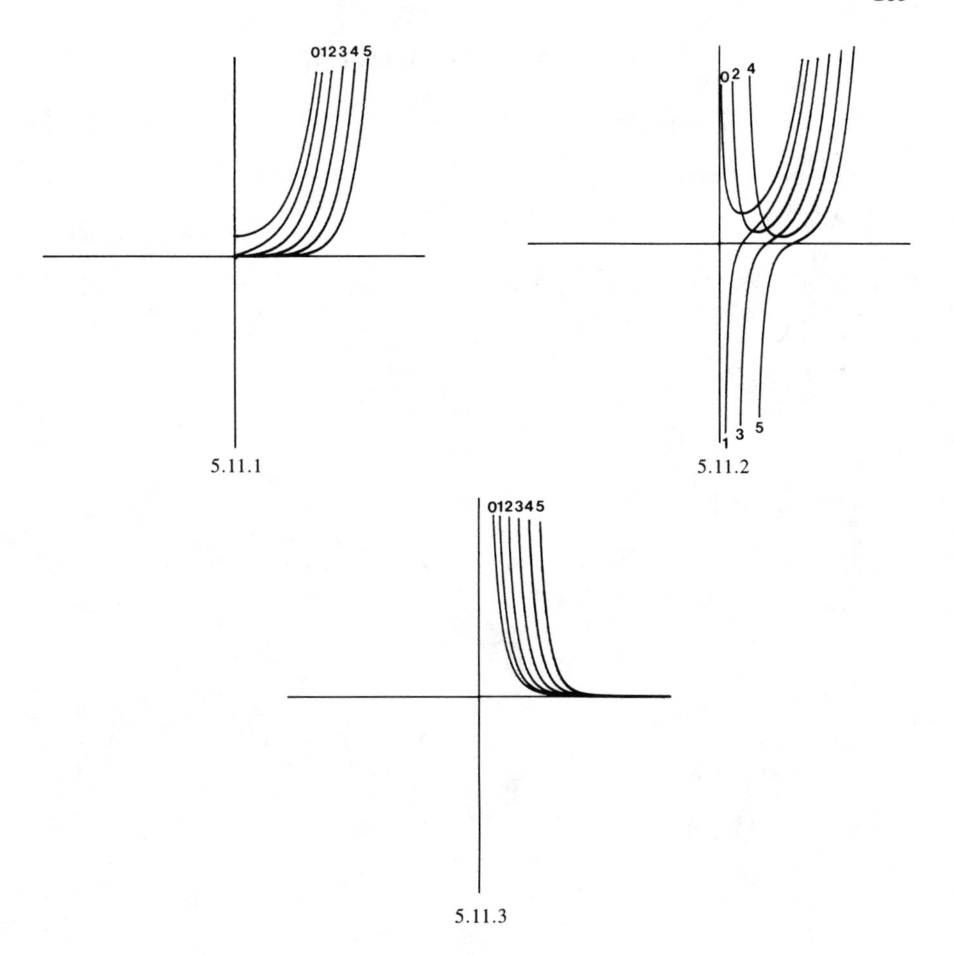

5.11.1

5.11.2

5.11.3

5.12. AIRY FUNCTIONS

5.12.1. Airy Function Ai(x)

Domain: $-\infty < x < \infty$

1. Ai(10x)

5.12.2. Airy Function Bi(x)

Domain: $-\infty < x < \infty$

1. Bi(10x)

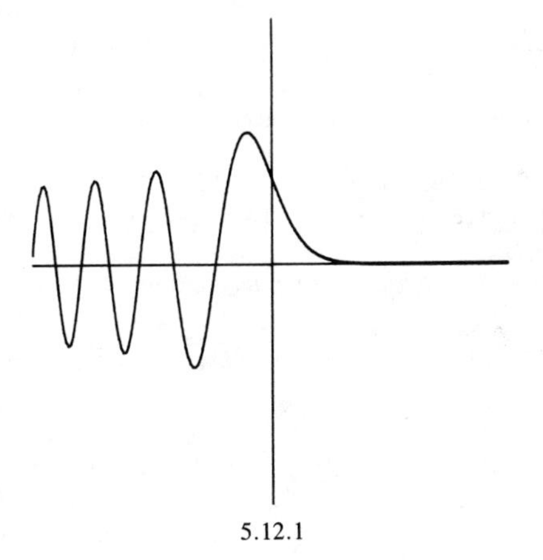

5.12.1

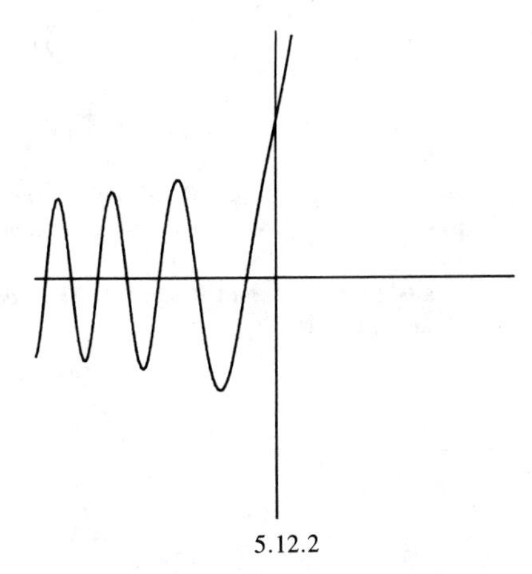

5.12.2

REFERENCES

1. **Abramowitz, M. and Stegun, I. A., Eds.,** Handbook of Mathematical Functions, National Bureau of Standards, U.S. Department of Commerce, Washington, D.C., 1964.
2. **Jahnke, E. and Emde, F.,** *Tables of Functions,* Dover Publications, New York, 1945.
3. **Beyer, W. H., Ed.,** *Handbook of Mathematical Sciences,* 6th ed., CRC Press, Boca Raton, FL, 1987.
4. **Gradshteyn, I. S. and Ryzhik, I. M.,** *Table of Integrals, Series, and Products,* Academic Press, Orlando, FL, 1973.

Chapter 6

SPECIAL FUNCTIONS IN PROBABILITY AND STATISTICS

6.1. DISCRETE PROBABILITY DENSITIES

The following discrete densities are plotted with the variable m on the x axis. Although a continuous line is plotted, the functions must be understood as discrete, having values only at interger m, the domain of which is listed in each case. The vertical scale is arbitrary, chosen to plot the density such that its maximum value is of uniform height for all plots. Thus, the scale may change among a series of plots for a given density function. A property common to all discrete densities is that the sum over all possible m must equal unity thus:

$$\sum_{m=m_1}^{m_2} P(m) = 1$$

where m_1 and m_2 are the minimum and maximum possible values of m.

6.1.1. Binomial $P(m|n,p) = \binom{n}{m}p^m(1 - p)^{n-m}$ where:

m = number of given outcomes
n = total number of trials
p = probability of a given outcome in one trial

1. n = 25, p = 0.25; m = 0,1,2,...,25
2. n = 25, p = 0.50; m = 0,1,2,...,25
3. n = 25, p = 0.75; m = 0,1,2,...,25
4. n = 10, p = 0.25; m = 0,1,2,...,10
5. n = 10, p = 0.50; m = 0,1,2,...,10
6. n = 10, p = 0.75; m = 0,1,2,...,10

6.1.2. Poisson $P(m|r) = e^{-r}r^m/m!$

m = number of events occurring in a given unit of time
r = mean rate (number of events per unit time)

1. r = 2; m = 0,1,2,...,25
2. r = 6; m = 0,1,2,...,25
3. r = 10; m = 0,1,2,...,25

6.1.3. Geometric $P(m|p) = p(1 - p)^{m-1}$

m = number of events
p = probability of a given event

1. p = 0.25; m = 1,2,3,...,10
2. p = 0.50; m = 1,2,3,...,10
3. p = 0.75; m = 1,2,3,...,10

6.1.4. Negative Binomial $P(m|n,p) = \binom{n+m-1}{m}p^n(1 - p)^m$

m = number of failures prior to nth success
n = number of successes
p = probability of a given event

1. n = 10, p = 0.25; m = 0,1,2,...,50
2. n = 10, p = 0.50; m = 0,1,2,...,50
3. n = 25, p = 0.50; m = 0,1,2,...,50
4. n = 25, p = 0.75; m = 0,1,2,...,50

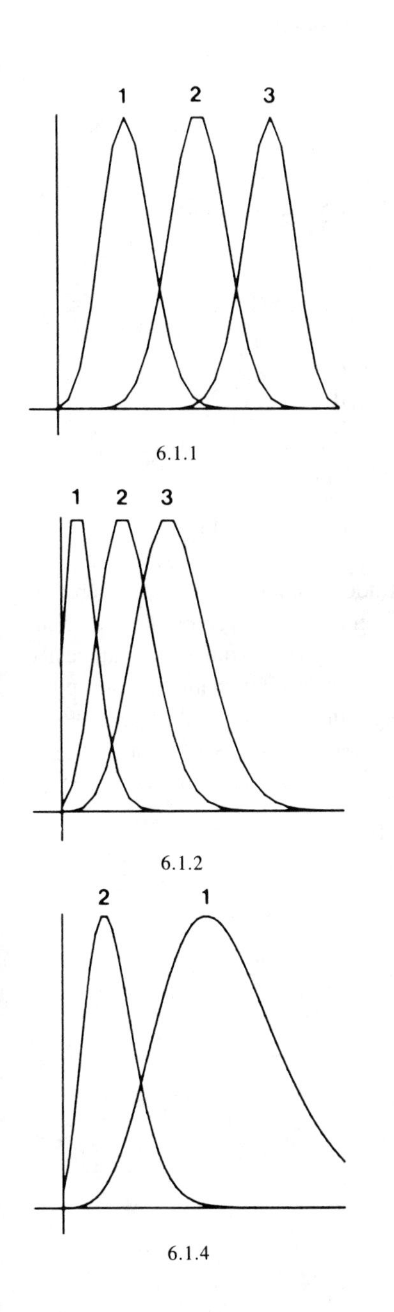

6.1.1

6.1.2

6.1.4

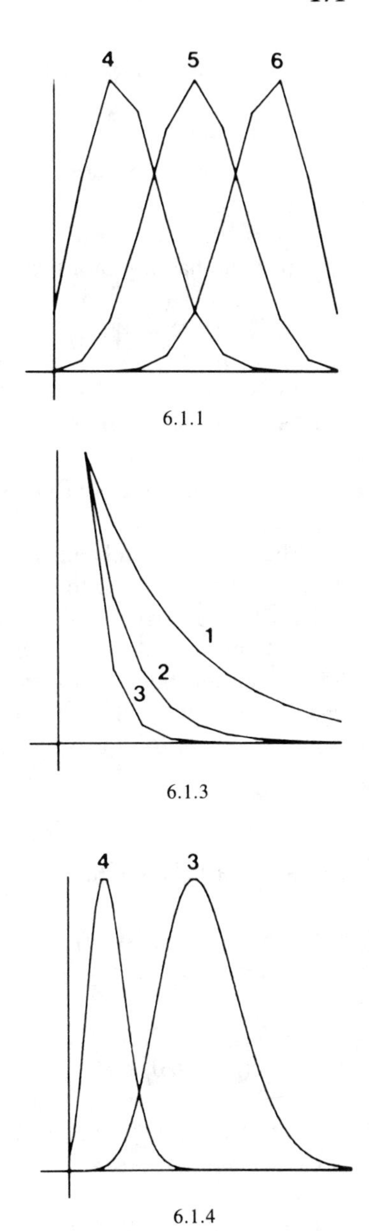

6.1.1

6.1.3

6.1.4

6.1.5. Hypergeometric $P(m|n,N,p) = \left(\frac{\binom{Np}{m}\binom{N(1-p)}{n-m}}{\binom{N}{n}}\right)$

m = number of items of a given type in a sample m = min(n, Np)
n = sample size
N = total number of items available (N > n)
p = probability of a given item type in total number N

1. n = 10, N = 40, p = 0.25; m = 0,1,2,...,10
2. n = 10, N = 40, p = 0.50; m = 0,1,2,...,10
3. n = 20, N = 40, p = 0.25; m = 0,1,2,...,20
4. n = 20, N = 40, p = 0.50; m = 0,1,2,...,20

6.2. CONTINUOUS PROBABILITY DENSITIES

The following probability densities are continuous functions, plotted such that the x-axis limits are −1 to +1 (with the actual domain of x given by the argument as listed). The range of y is arbitrary, selected only to plot the function in an easily viewable manner. As for the discrete densities, the scale may change among plots for a given function. A property common to all continuous probability densities is that the integral equals unity thus:

$$\int_a^b P(x)\,dx = 1$$

where *a* and *b* are the limits of the particular density function.

6.2.1. Normal (Gaussian) $P(x) = [1/(2\pi)^{1/2}b]\exp\{-[(x - a)/b]^2/2\}$
Domain: $-\infty < x < \infty$

1. P(5x): a = 0, b = 1
2. P(5x): a = 0, b = 2

6.2.2. Cauchy $P(x) = (1/\pi b)\{1 + [(x - a)/b]^2\}^{-1}$
Domain: $-\infty < x < \infty$

1. P(5x): a = 0, b = 1
2. P(5x): a = 0, b = 2

6.2.3. Exponential $P(x) = (1/b)\exp[-(x - a)/b]$
Domain: $a < x < \infty$

1. P(5x): a = 0, b = 1
2. P(5x): a = 0, b = 2

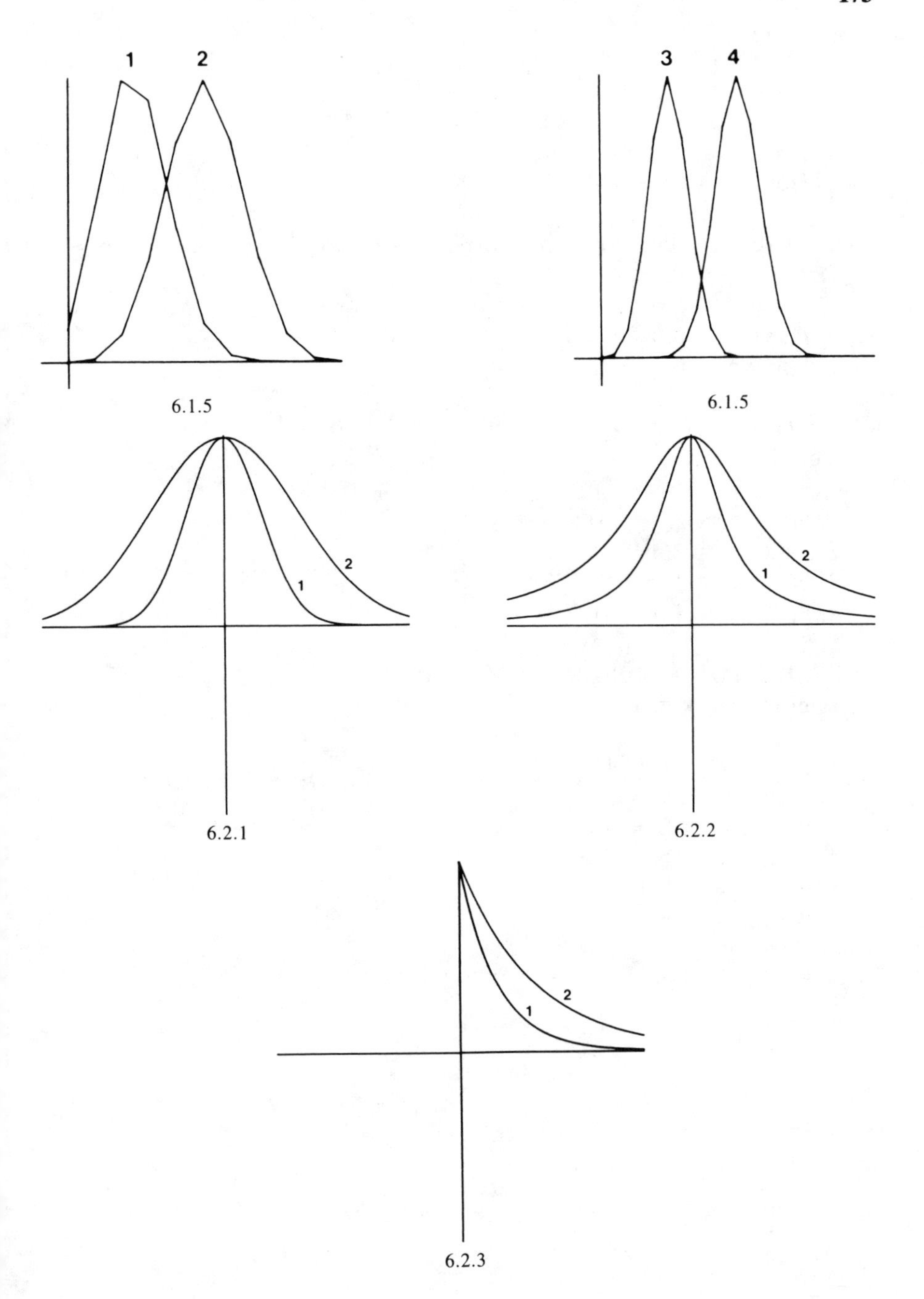

6.1.5

6.1.5

6.2.1

6.2.2

6.2.3

6.2.4. Laplace $P(x) = (1/2b)\exp[-|(x - a)/b|]$
Domain: $-\infty < x < \infty$

1. P(5x): a = 0, b = 1
2. P(5x): a = 0, b = 2

6.2.5. Extreme Value $P(x) = (1/b)\exp\{-|(x - a)/b| - \exp[-|(x - a)/b|]\}$
Domain: $-\infty < x < \infty$

1. P(5x): a = 0, b = 1
2. P(5x): a = 0, b = 2

6.2.6. Gamma $P(x) = [1/\Gamma(a)b^a]x^{a-1}e^{-x/b}$ (a,b > 0)
Domain: $x > 0$

1. P(5x): a = 2, b = 0.5
2. P(5x): a = 2, b = 1.0
3. P(5x): a = 3, b = 0.5
4. P(5x): a = 3, b = 1.0

6.2.7. Beta $P(x) = [1/B(a,b)]x^{a-1}(1 - x)^{b-1}$
Domain: $0 < x < 1$

1. P(x): a = 2, b = 1
2. P(x): a = 2, b = 2
3. P(x): a = 2, b = 4
4. P(x): a = 3, b = 2
5. P(x): a = 3, b = 3
6. P(x): a = 3, b = 6

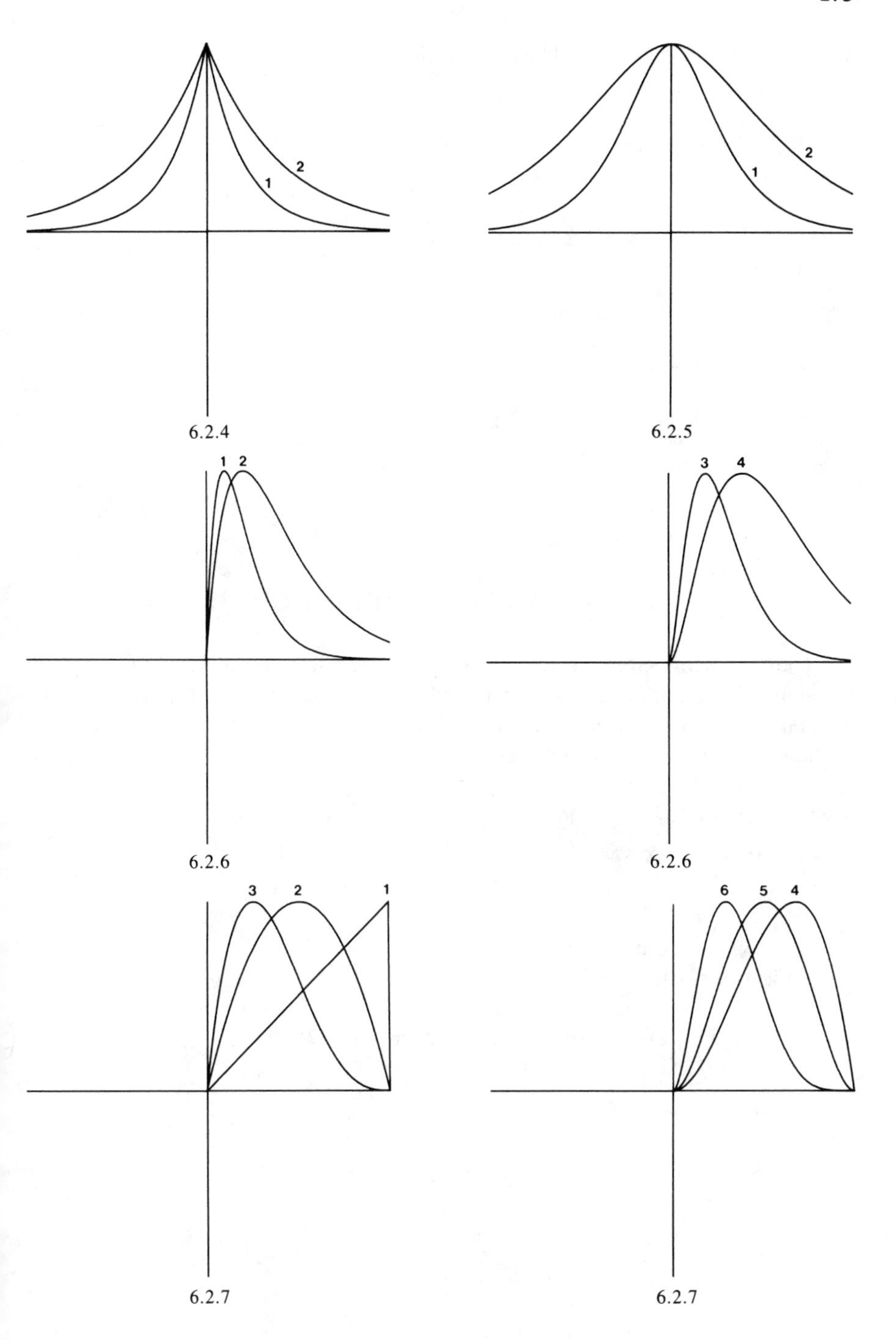

6.2.4 6.2.5

6.2.6 6.2.6

6.2.7 6.2.7

6.2.8. Log-Normal $P(x) = [1/(2\pi)^{1/2}b]\exp\{-[(\ln x - a)/b]^2/2\}$
Domain: $x > 0$

1. P(5x): a = 0, b = 0.5
2. P(5x): a = 0, b = 1.0

6.2.9. Rayleigh $P(x) = (1/a^2)x \cdot \exp[-(x/a)^2/2]$
Domain: $x > 0$

1. P(5x): a = 1
2. P(5x): a = 2

6.2.10. Maxwell $P(x) = (4/\pi^{1/2}a^3)x^2\exp(-x^2/a^2)$
Domain: $x > 0$

1. P(5x): a = 1
2. P(5x): a = 2

6.3. SAMPLING DISTRIBUTIONS

The following sampling distributions are expressed as integrals of a density function. By definition, at the upper limit the integral equals unity; therefore, the distributions are plotted such that the maximum is always unity. The actual domain of the sampling variable is as listed.

6.3.1. Normal Distribution $P(X) = [1/(2\pi)^{1/2}b]\int_{-\infty}^{x} \exp\{-[(x - a)/b]^2/2\}dx$
Domain: $-\infty < X < \infty$

1. P(5X): a = 0, b = 0.5
2. P(5X): a = 0, b = 1.0
3. P(5X): a = 0, b = 2.0

6.3.2. Student's t Distribution $P(t|n) = [n^{1/2}B(1/2,n/2)]^{-1}\int_{-\infty}^{t}$
$[1 + (x^2/n)]^{-(n+1)/2}dx$
Domain: $-\infty < t < \infty$

1. P(5t): n = 1
2. P(5t): n = 5
3. P(5t): n = 20

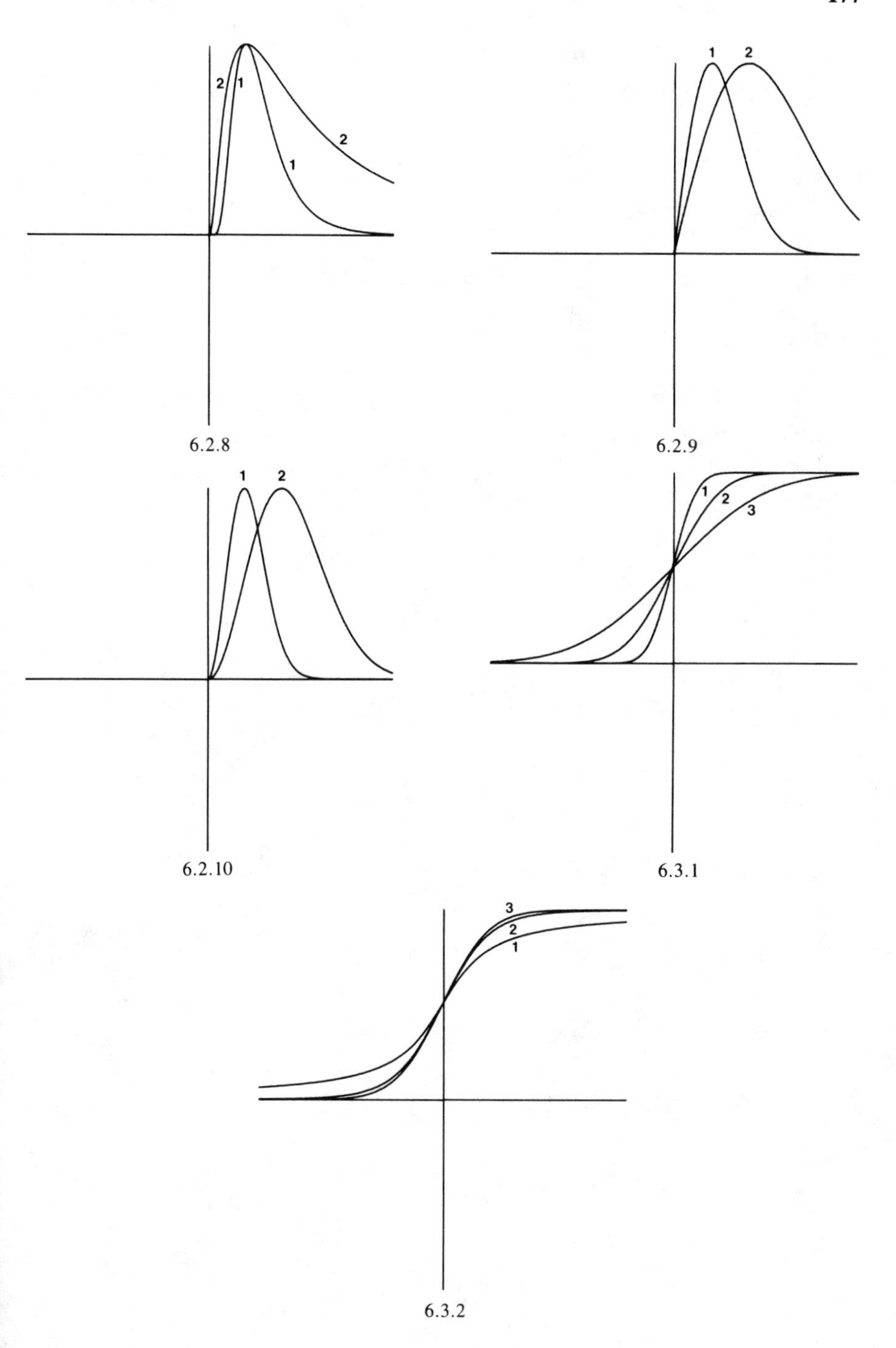

6.2.8

6.2.9

6.2.10

6.3.1

6.3.2

6.3.3. Chi-Square Distribution $P(\chi^2|n) = [2^{n/2}\Gamma(n/2)]^{-1}\int_0^{\chi^2} x^{(n-2)/2}e^{-x/2}dx$

Domain: $\chi^2 > 0$

1. $P(25\chi^2)$: $n = 2$
2. $P(25\chi^2)$: $n = 6$
3. $P(25\chi^2)$: $n = 16$

6.3.4. F Distribution $P(F|m,n) = [m^{m/2}n^{n/2}/B(m/2,n/2)]\int_0^F x^{(m-2)/2}(n + mx)^{-(m+n)/2}dx$

Domain: $F > 0$

1. $P(5F)$: $m = 2$, $n = 10$
2. $P(5F)$: $m = 6$, $n = 10$
3. $P(5F)$: $m = 20$, $n = 20$

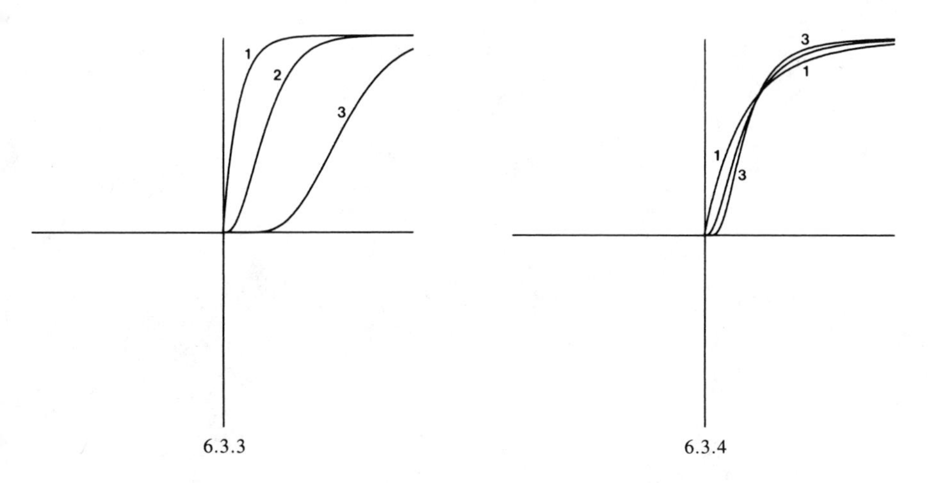
1
2
3
6.3.3
3
1
1
3
6.3.4

Chapter 7

MISCELLANEOUS CURVES

7.1. CATASTROPHE CURVES

The following curves can be found in most works on catastrophe theory (for example, Poston and Stewart[1]) but are only a limited sampling of curves commonly associated with this theory. The curves shown here are merely representative, and forms for a wider range of parameters can be found in the above reference.

7.1.1. Cusp Catastrophe

$y = -(ax^2)^{1/3}$

1. $a = 4.0$
2. $a = 8.0$

7.1.2. Butterfly Catastrophe

$x = c(8at^3 + 24t^5)$
$y = c(-6at^2 - 15t^4)$

1. $a = -5.0$, $c = 0.03$; $-1.455 < t < 1.455$
2. $a = -7.0$, $c = 0.03$; $-1.629 < t < 1.629$

7.1.3. Swallowtail Catastrophe

$x = c(-2at - 4t^3)$
$y = c(at^2 + 3t^4)$

1. $a = -1.0$, $c = 0.5$; $-1 < t < 1$
2. $a = -2.0$, $c = 0.5$; $-2 < t < 2$

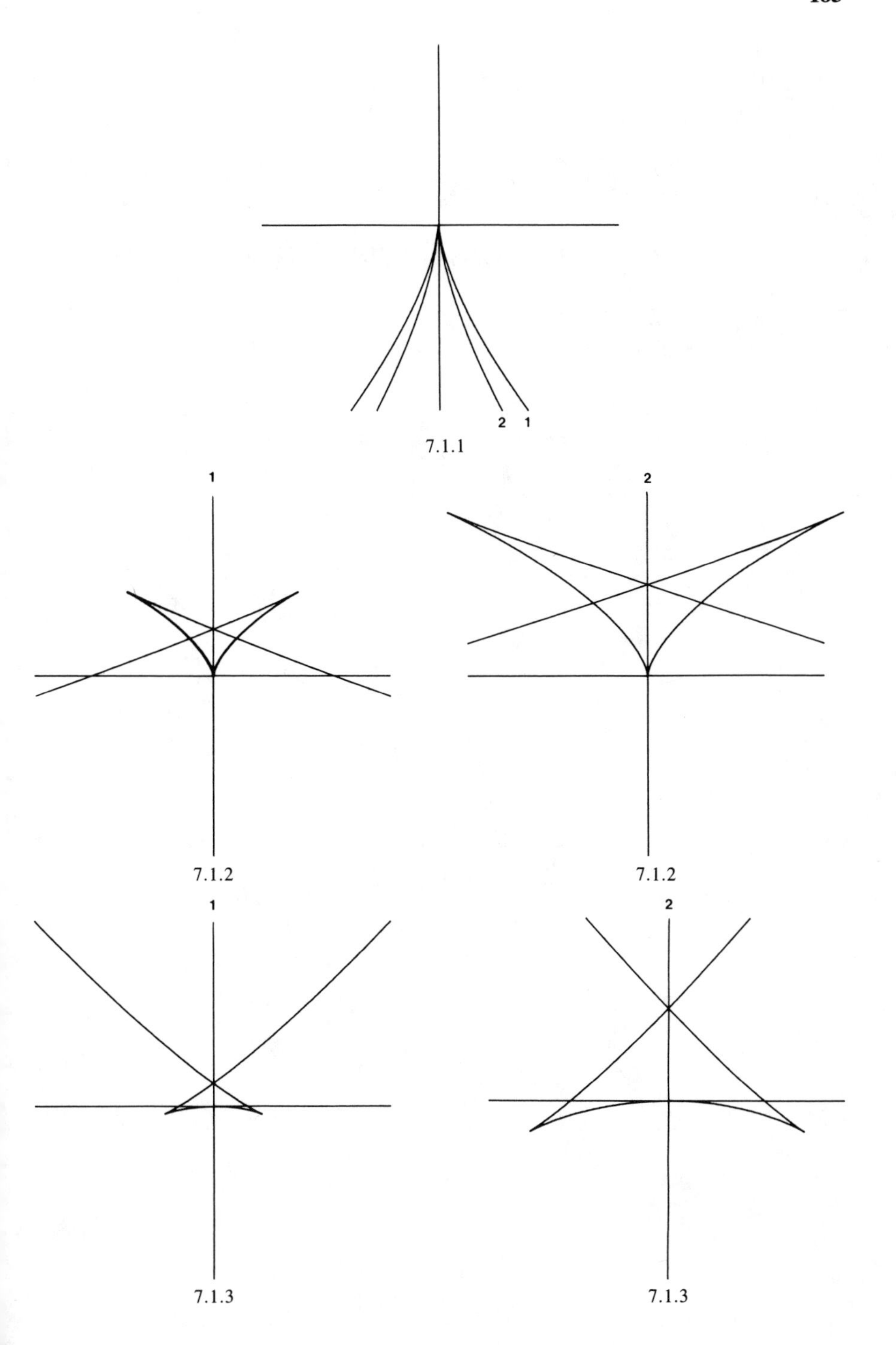

7.1.1

7.1.2

7.1.2

7.1.3

7.1.3

7.1.4. n-Roll Mill

$x^n - \binom{n}{2}x^{n-2}y^2 + \binom{n}{4}x^{n-4}y^4 - \ldots = a$

1. $n = 2, a = 0.10$
2. $n = 3, a = 0.03$
3. $n = 4, a = 0.01$

7.2. VARIATIONS ON SINE CURVES

7.2.1. Modulated Sine Wave

$y = c \cdot \sin(2\pi ax)\sin(2\pi bx)$

1. $a = 10., b = 0.5, c = 0.50$
2. $a = 10., b = 1.0, c = 0.50$

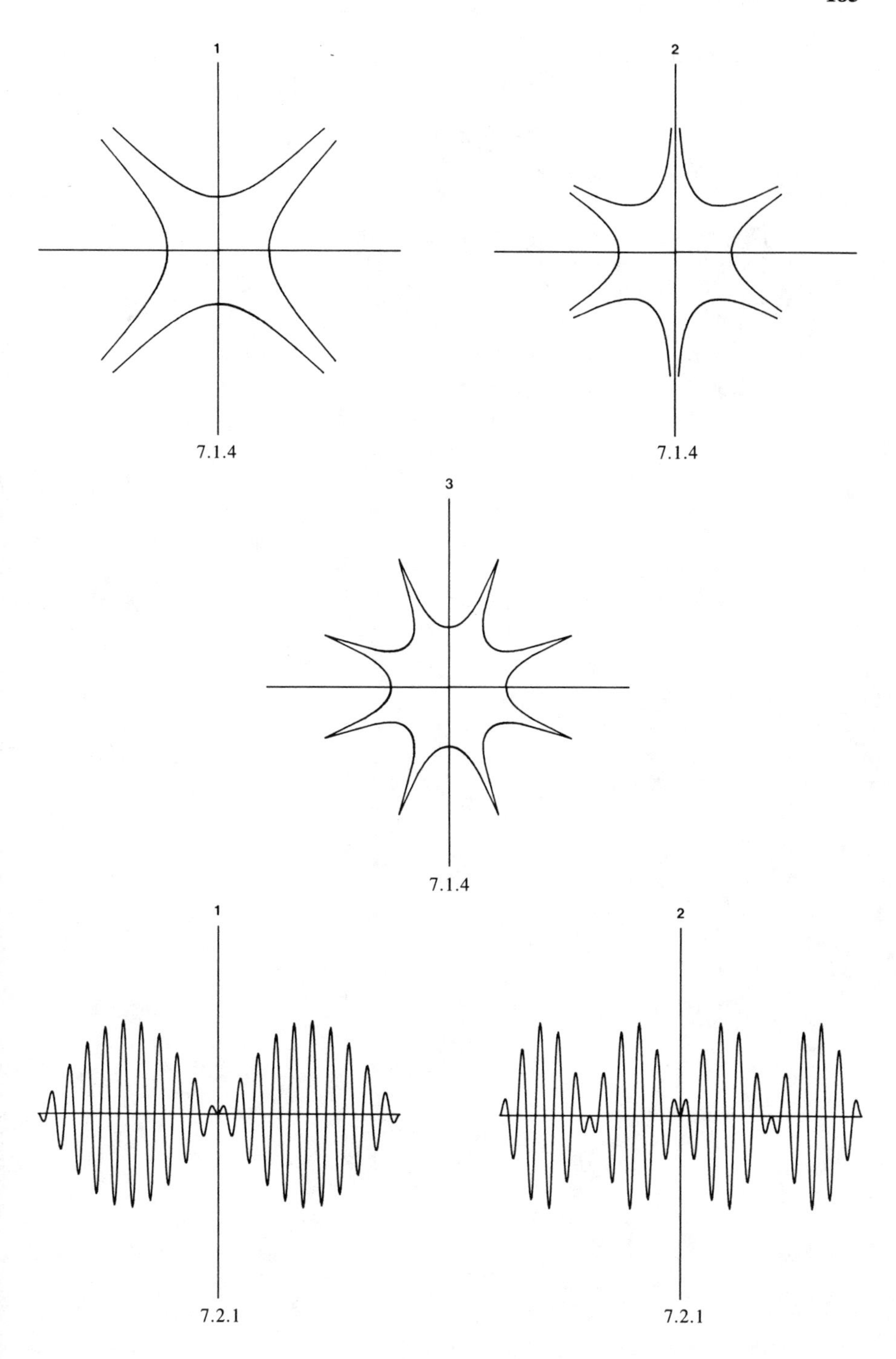

7.1.4 7.1.4

7.1.4

7.2.1 7.2.1

7.2.2. Sweep Signal (Linear)

$y = c \cdot \sin(\{\pi d/[b - a]\}\{[(b - a)(x/d) + a]^2 - a^2\})$

1. $a = 10.$, $b = 20.$, $c = 0.5$, $d = 1.0$; $0 < x < 1$
2. $a = 5.$, $b = 25.$, $c = 0.5$, $d = 1.0$; $0 < x < 1$

7.2.3. Lissajous Curves (also called Bowditch curves)

$x = \sin(at + b\pi)$ (a is rational)

$y = \sin(t)$

1. $a = 1/2$, $b = 0.0$; $0 < t < 2(2\pi)$
2. $a = 1/3$, $b = 0.0$; $0 < t < 3(2\pi)$
3. $a = 1/4$, $b = 0.0$; $0 < t < 4(2\pi)$
4. $a = 1/5$, $b = 0.0$; $0 < t < 5(2\pi)$

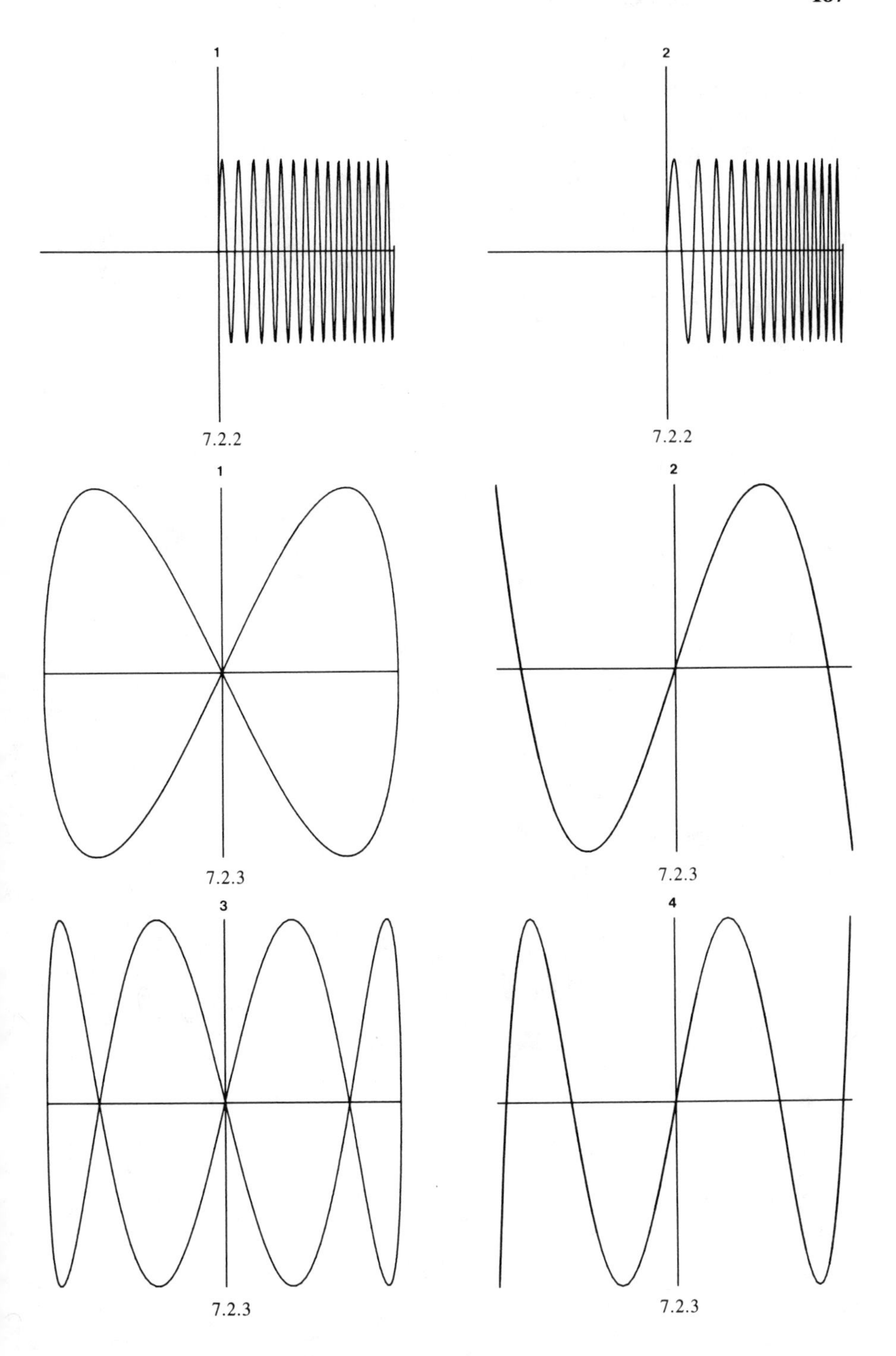

7.2.2

7.2.2

7.2.3

7.2.3

7.2.3

7.2.3

5. $a = 2/3$, $b = 0.0$; $0 < t < 3(2\pi)$
6. $a = 3/4$, $b = 0.0$; $0 < t < 4(2\pi)$
7. $a = 1/2$, $b = 0.2$; $0 < t < 2(2\pi)$
8. $a = 1/3$, $b = 0.2$; $0 < t < 3(2\pi)$
9. $a = 1/4$, $b = 0.2$; $0 < t < 4(2\pi)$
10. $a = 1/5$, $b = 0.2$; $0 < t < 5(2\pi)$

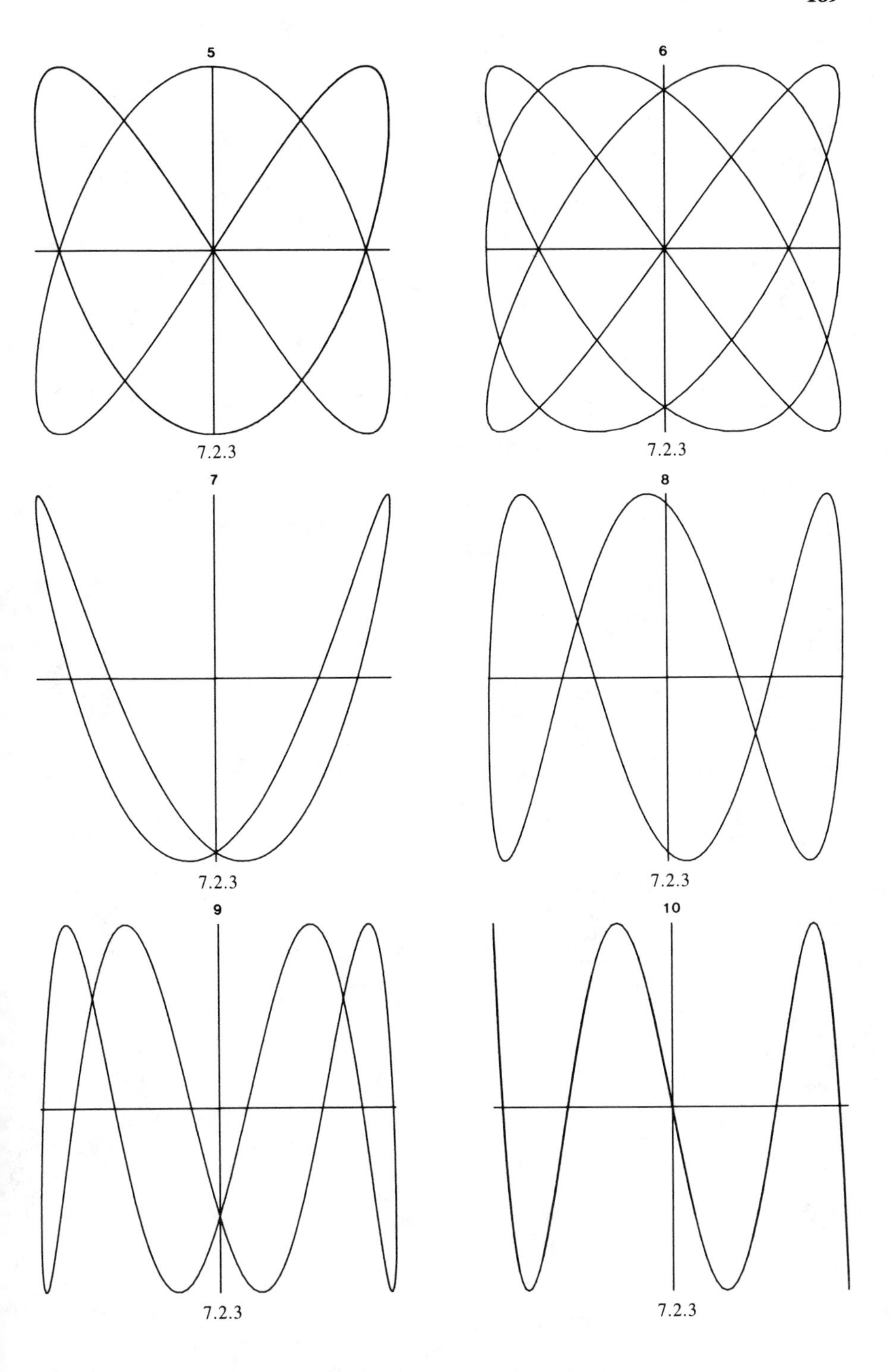

7.2.3

7.2.3

7.2.3

7.2.3

7.2.3

7.2.3

11. $a = 2/3$, $b = 0.2$; $0 < t < 3(2\pi)$
12. $a = 3/4$, $b = 0.2$; $0 < t < 4(2\pi)$
13. $a = 1/2$, $b = 0.4$; $0 < t < 2(2\pi)$
14. $a = 1/3$, $b = 0.4$; $0 < t < 3(2\pi)$
15. $a = 1/4$, $b = 0.4$; $0 < t < 4(2\pi)$
16. $a = 1/5$, $b = 0.4$; $0 < t < 5(2\pi)$

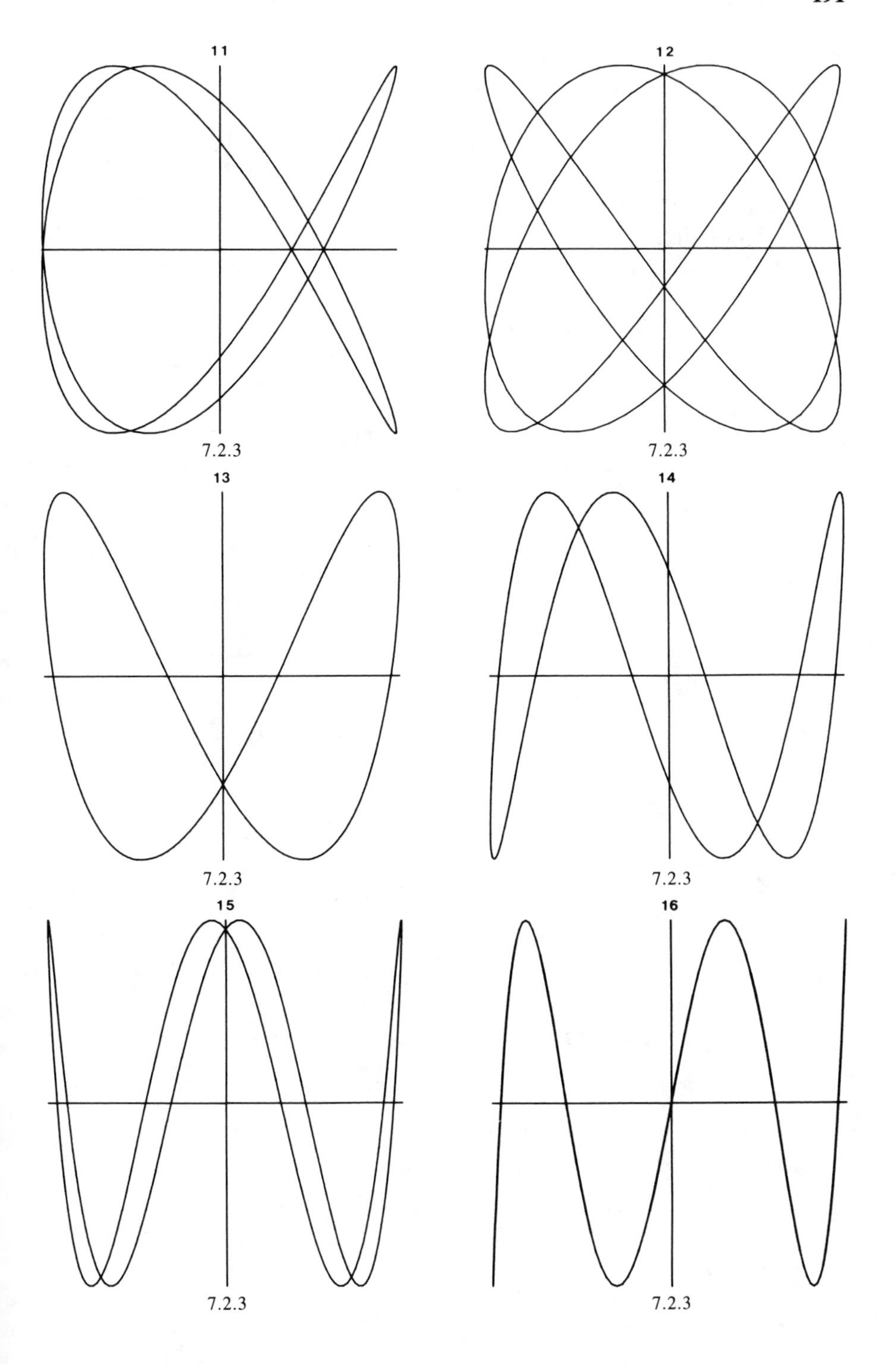

7.2.3

7.2.3

7.2.3

7.2.3

7.2.3

7.2.3

17. $a = 2/3$, $b = 0.4$; $0 < t < 3(2\pi)$
18. $a = 3/4$, $b = 0.4$; $0 < t < 4(2\pi)$

7.3. CYCLOIDS

7.3.1. Ordinary Cycloid

$x = a[t - \sin(t)]$
$y = c[1 - \cos(t)]$

1.$a = 0.1$, $c = 0.5$; $-10 < t < 10$

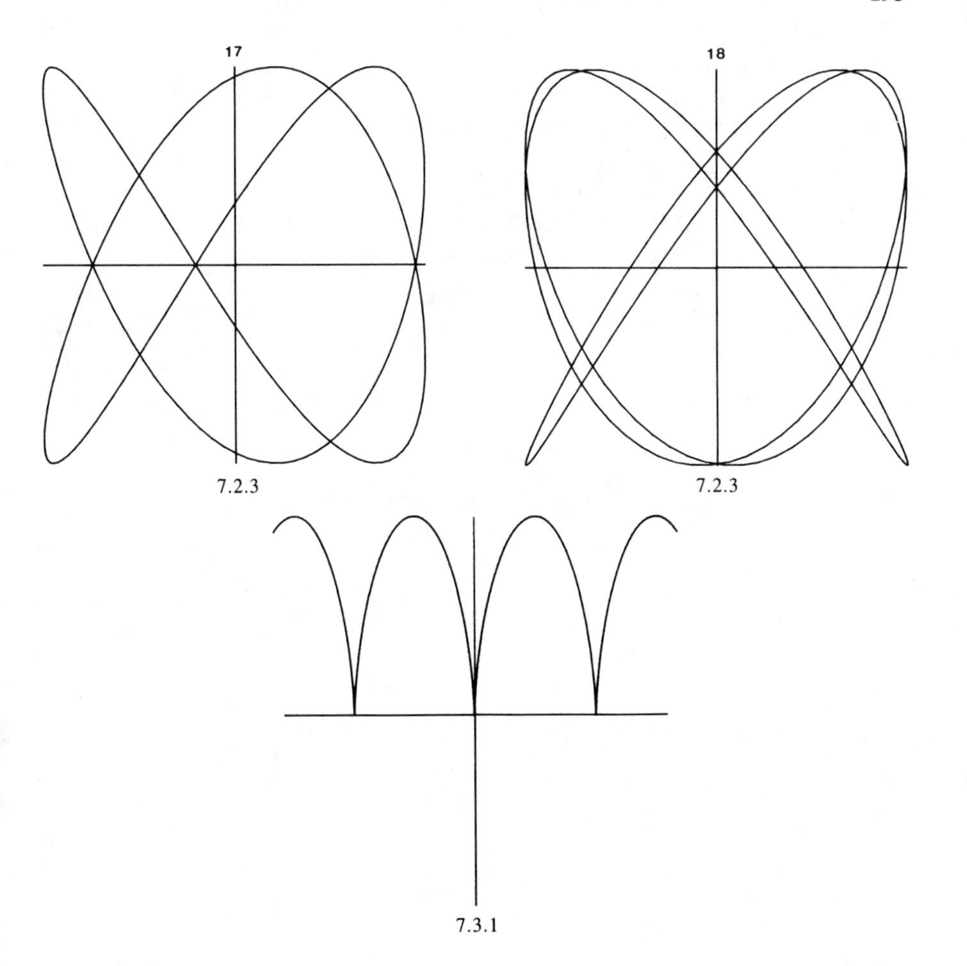

7.2.3 7.2.3

7.3.1

7.3.2. Prolate Cycloid

$x = a[t - b \cdot \sin(t)]$

$y = c[1 - d \cdot \cos(t)]$

1. $a = 0.1$, $b = 2.0$, $c = 0.2500$, $d = 2.0$; $-10 < t < 10$
2. $a = 0.1$, $b = 4.0$, $c = 0.1250$, $d = 4.0$; $-10 < t < 10$
3. $a = 0.1$, $b = 8.0$, $c = 0.0625$, $d = 8.0$; $-10 < t < 10$

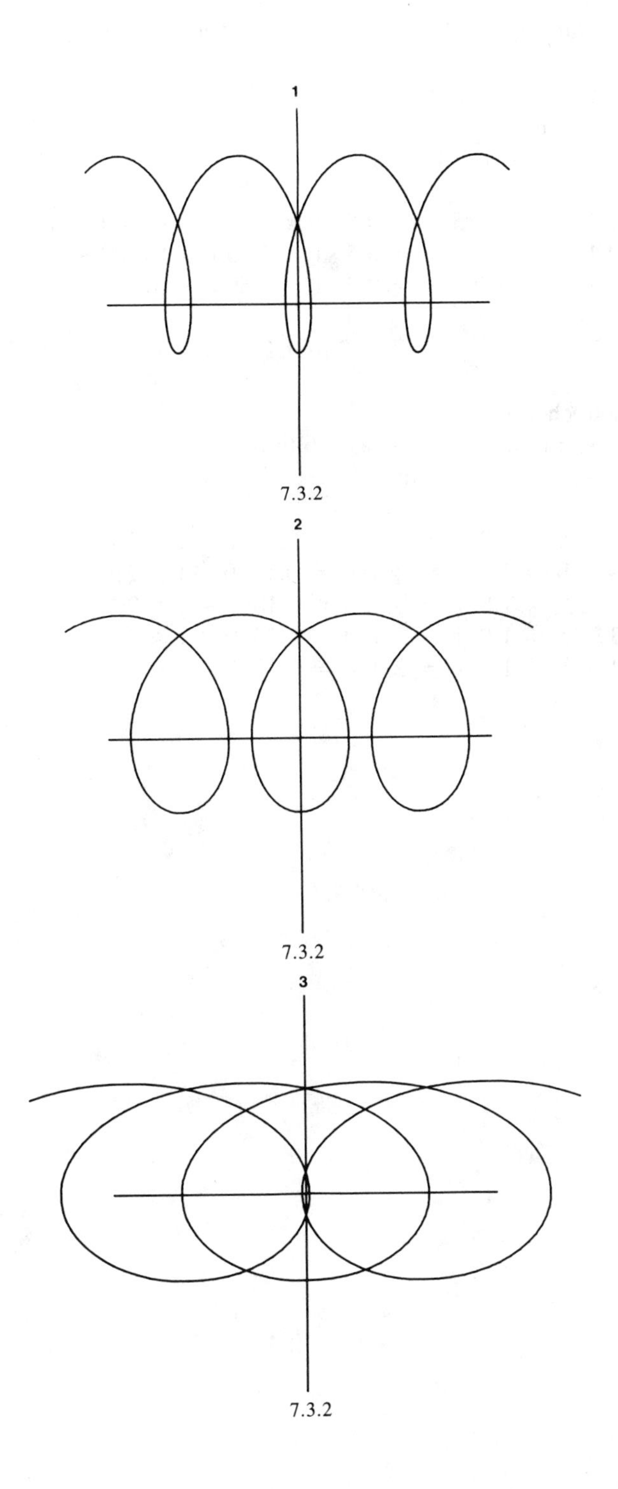

7.3.2

7.3.2

7.3.2

7.3.3. Curtate Cycloid

$x = a[t - b \cdot \sin(t)]$
$y = c[1 - d \cdot \cos(t)]$

1. $a = 0.1, b = 0.125, c = 0.5, d = 0.125; -10 < t < 10$
2. $a = 0.1, b = 0.250, c = 0.5, d = 0.250; -10 < t < 10$
3. $a = 0.1, b = 0.500, c = 0.5, d = 0.500; -10 < t < 10$

7.4. TROCHOIDS

7.4.1. Hypotrochoid

$x = d\{(a - b)\cos(t) + c \cdot \cos[(a - b)t/b]\}$
$y = d\{(a - b)\sin(t) - c \cdot \sin[(a - b)t/b]\}$

1. $a = 3.0, b = 1.0, c = 3.0, d = 0.15; 0 < t < 2\pi$
2. $a = 4.0, b = 1.0, c = 3.0, d = 0.15; 0 < t < 2\pi$
3. $a = 5.0, b = 1.0, c = 3.0, d = 0.10; 0 < t < 2\pi$
4. $a = 3.0, b = 1.0, c = 2.0, d = 0.25; 0 < t < 2\pi$
5. $a = 4.0, b = 1.0, c = 2.0, d = 0.15; 0 < t < 2\pi$

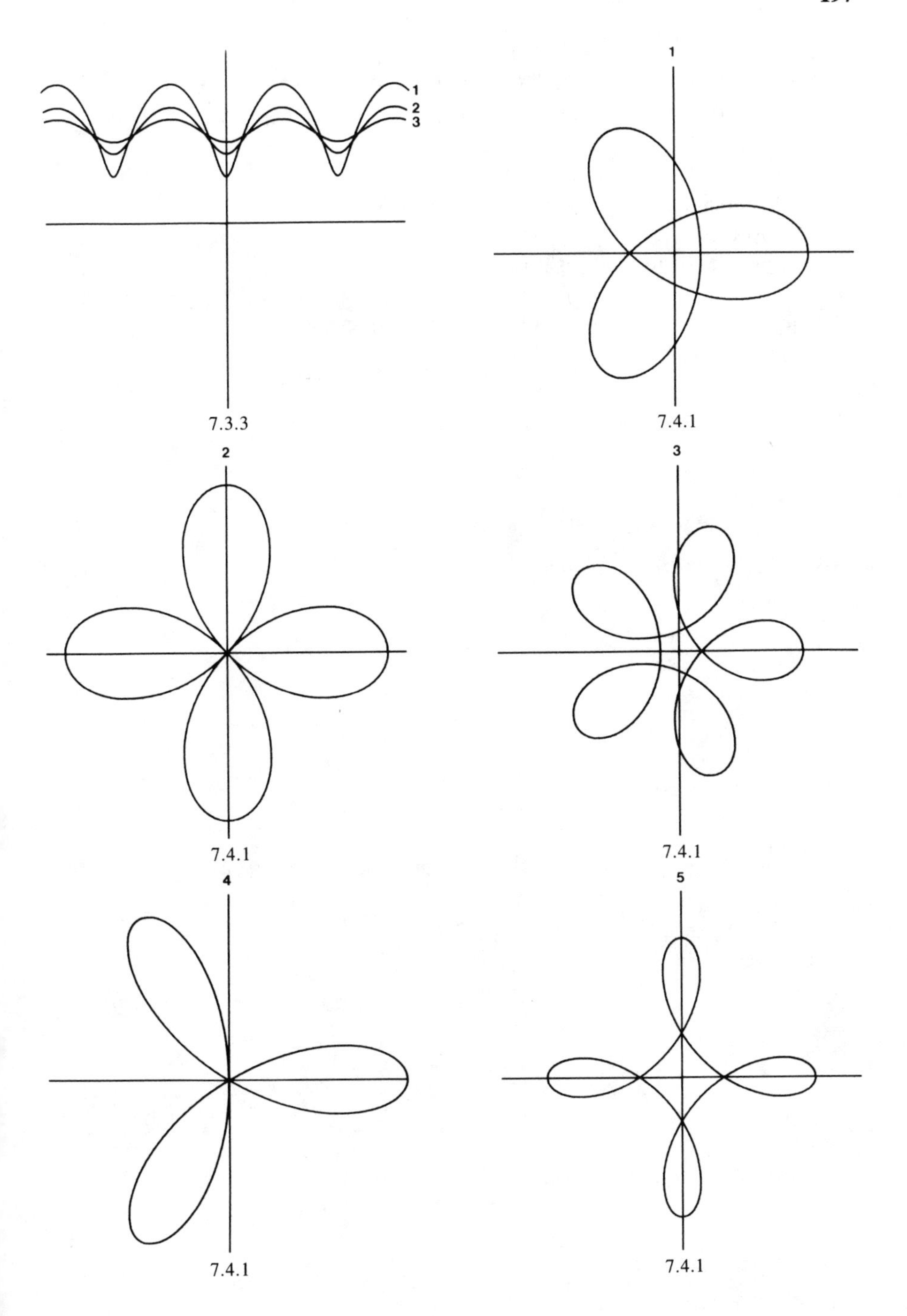

7.3.3

7.4.1

7.4.1

7.4.1

7.4.1

7.4.1

6. $a = 5.0, b = 1,0, c = 2.0, d = 0.15; 0 < t < 2\pi$
7. $a = 3.0, b = 1.0, c = 1.0, d = 0.25; 0 < t < 2\pi$ (''deltoid'')
8. $a = 4.0, b = 1.0, c = 1.0, d = 0.25; 0 < t < 2\pi$ (''astroid'')
9. $a = 5.0, b = 1.0, c = 1.0, d = 0.20; 0 < t < 2\pi$

7.4.2. Epitrochoid

$$x = d\{(a + b)\cos(t) - c\cdot\cos[(a + b)t/b]\}$$
$$y = d\{(a + b)\sin(t) - c\cdot\sin[(a + b)t/b]\}$$

1. $a = 3.0, b = 1,0, c = 3.0, d = 0.15; 0 < t < 2\pi$
2. $a = 4.0, b = 1.0, c = 3.0, d = 0.15; 0 < t < 2\pi$

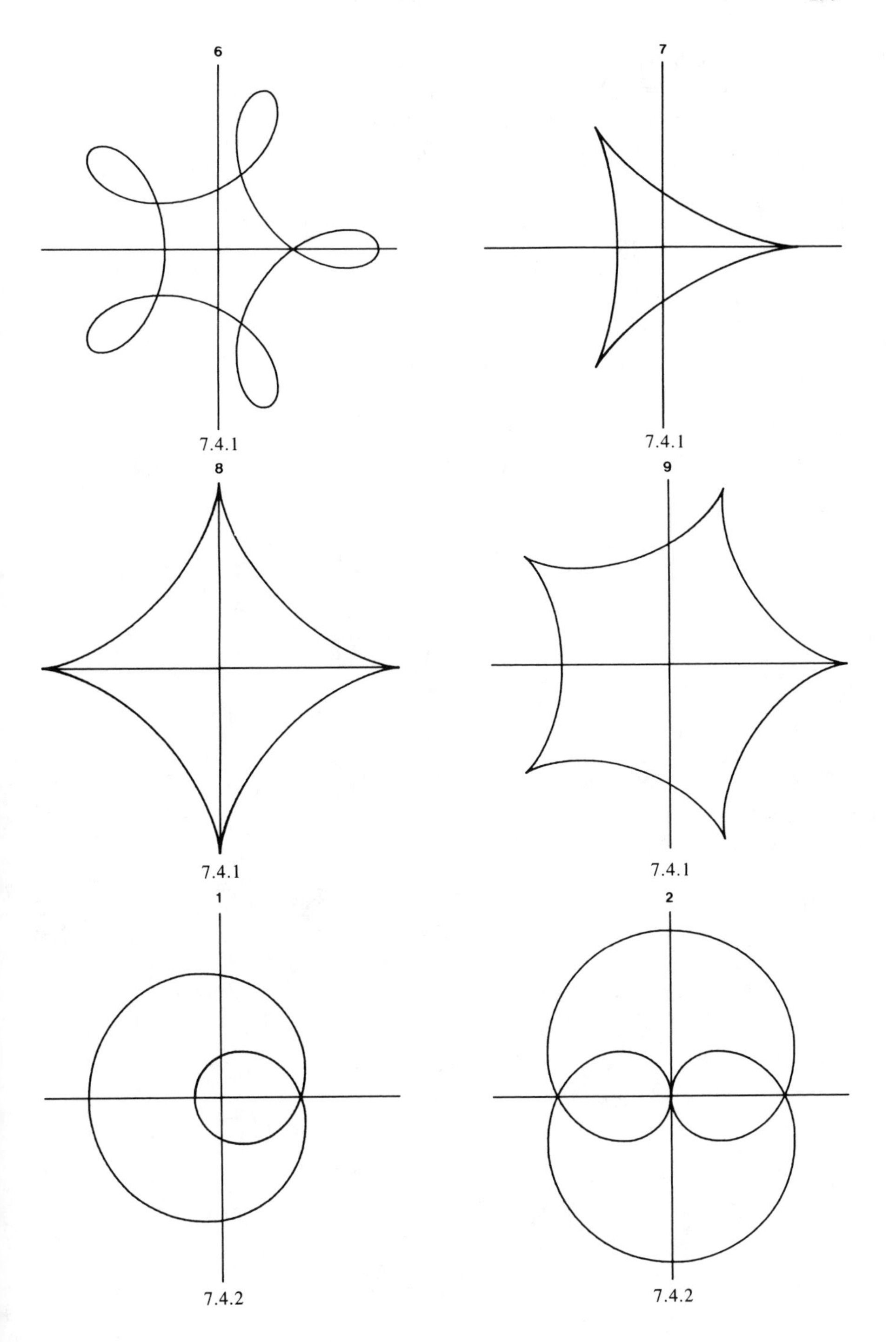

7.4.1

7.4.1

7.4.1

7.4.1

7.4.2

7.4.2

3. $a = 5.0$, $b = 1.0$, $c = 3.0$, $d = 0.10$; $0 < t < 2\pi$
4. $a = 3.0$, $b = 1.0$, $c = 2.0$, $d = 0.25$; $0 < t < 2\pi$
5. $a = 4.0$, $b = 1.0$, $c = 2.0$, $d = 0.15$; $0 < t < 2\pi$
6. $a = 5.0$, $b = 1.0$, $c = 2.0$, $d = 0.15$; $0 < t < 2\pi$

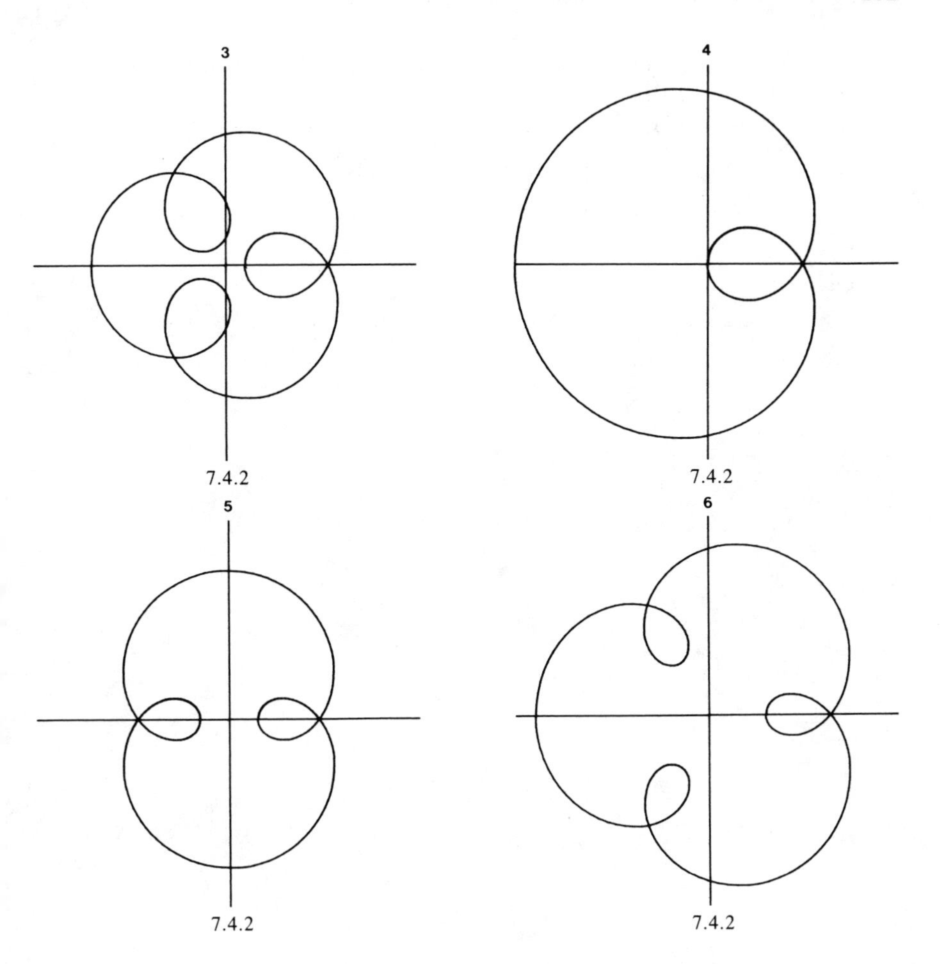

7.4.2

7.4.2

7.4.2

7.4.2

7. $a = 3.0$, $b = 1.0$, $c = 1.0$, $d = 0.25$; $0 < t < 2\pi$ (''nephroid'')
8. $a = 4.0$, $b = 1.0$, $c = 1.0$, $d = 0.25$; $0 < t < 2\pi$
9. $a = 5.0$, $b = 1.0$, $c = 1.0$, $d = 0.20$; $0 < t < 2\pi$

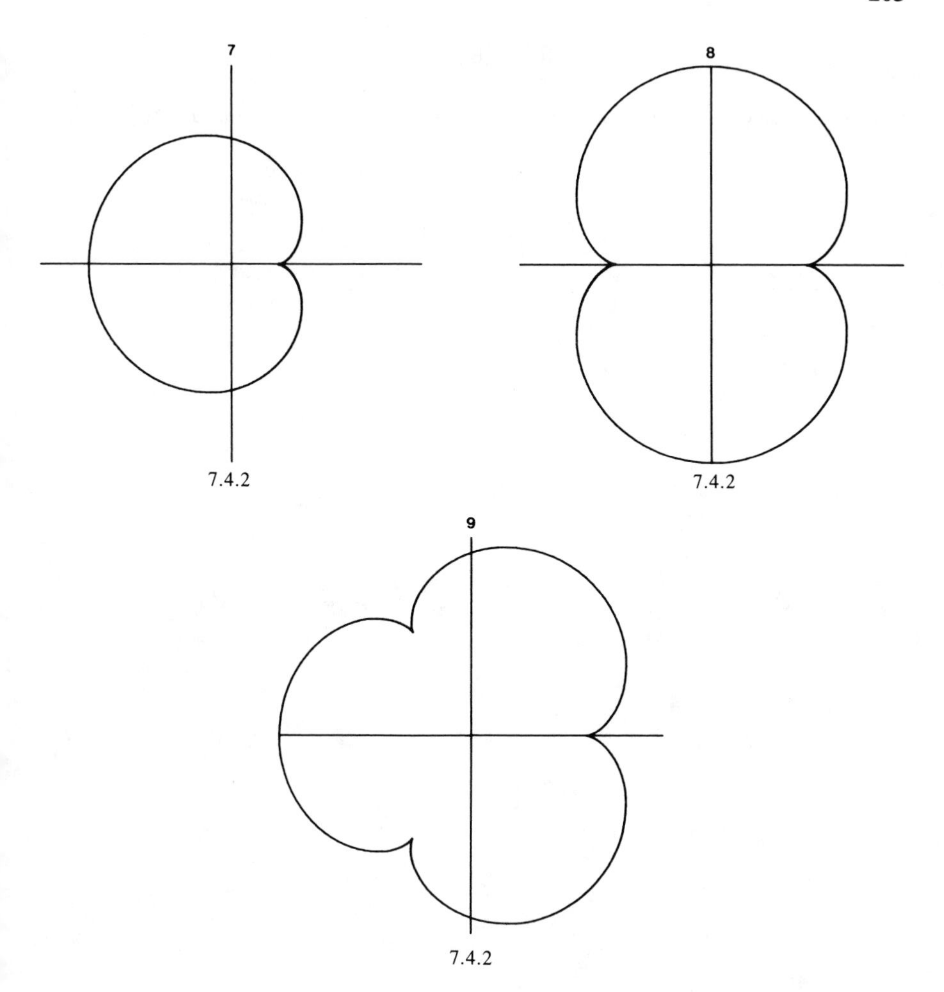

7.4.2

7.4.2

7.4.2

REFERENCE

1. **Poston, T. and Stewart, I.,** *Catastrophe Theory and Its Applications,* Pitman Publishing, London, 1978.

Chapter 8

THREE-DIMENSIONAL CURVES

As opposed to curves which lie wholly in a plane (called "plane curves"), certain curves occupy three dimensions (called "skew curves"). All three-dimensional curves must necessarily be expressed in parametric form:

$$x = f(t)$$
$$y = g(t)$$
$$z = h(t)$$

Because there are innumerable variations of the functions f, g, and h, three-dimensional curves can assume a wide variety in appearance. Only those curves having some accepted significance and use are illustrated here. Many interesting and useful three-dimensional curves can be generated simply by adding a *z* variation of some sort to the curves given in the previous chapters, after they are put into parametric form.

The curves in this chapter are plotted as points (x_p,y_p) projected on a plane which is normal to the vector between the origin at (0,0,0) and the viewpoint at infinity and which passes through the origin. The projection used is the "orthographic parallel" one (see Foley and VanDam[1] for a full treatment of projections), which is given by the transformations:

$$x_p = -x\cdot\sin\theta + y\cdot\cos\theta$$
$$y_p = -x\cdot\cos\theta\cos\phi - y\cdot\sin\theta\cos\phi + z\cdot\sin\phi$$

where (x,y,z) are the coordinates of the point on the curve prior to projection and (θ,ϕ) are the angles in spherical coordinates (see Section 1.3) of the vector normal to the projection plane. The three axes are plotted with dashed lines between the limits of -1.0 and $+1.0$, with the positive z axis up.

8.1. HELICAL CURVES

8.1.1. Circular Helix

$x = a \cdot \sin t$
$y = a \cdot \cos t$
$z = t/2\pi c$

1. $a = 0.5,\ c = 5.0;\ 0 < t < 10\pi;\ \theta = 315,\ \phi = 60$

8.1.2. Conical Helix

$x = az \cdot \sin t$
$y = az \cdot \cos t$
$z = t/2\pi c$

1. $a = 0.5,\ c = 5.0;\ 0 < t < 10\pi;\ \theta = 315,\ \phi = 60$

8.1.3. Elliptic Cylindrical Helix

$x = a \cdot \sin t$
$y = b \cdot \cos t$
$z = t/2\pi c$

1. $a = 0.3,\ b = 1.0,\ c = 5.0;\ 0 < t < 10\pi;\ \theta = 315,\ \phi = 60$

8.2. SINE WAVES IN THREE DIMENSIONS

8.2.1. Sine Wave on Sphere

$x = b[1 - c^2\cos^2(at)]^{1/2}\cos(t)$
$y = b[1 - c^2\cos^2(at)]^{1/2}\sin(t)$
$z = c \cdot \cos(at)$

1. $a = 10.,\ b = 1.0,\ c = 0.30;\ 0 < t < 2\pi;\ \theta = 315,\ \phi = 60$

8.2.2. Sine Wave on Hyperboloid of One Sheet

$x = b[1 + c^2\cos^2(at)]^{1/2}\cos(t)$
$y = b[1 + c^2\cos^2(at)]^{1/2}\sin(t)$
$z = c \cdot \cos(at)$

1. $a = 10.,\ b = 0.8,\ c = 0.30;\ 0 < t < 2\pi;\ \theta = 315,\ \phi = 60$

8.2.3. Sine Wave on Cone

$x = c[1 + \cos(at)]\cos(t)$
$y = c[1 + \cos(at)]\sin(t)$
$z = c[1 + \cos(at)]$

1. $a = 10.,\ c = 0.5;\ 0 < t < 2\pi;\ \theta = 315,\ \phi = 60$

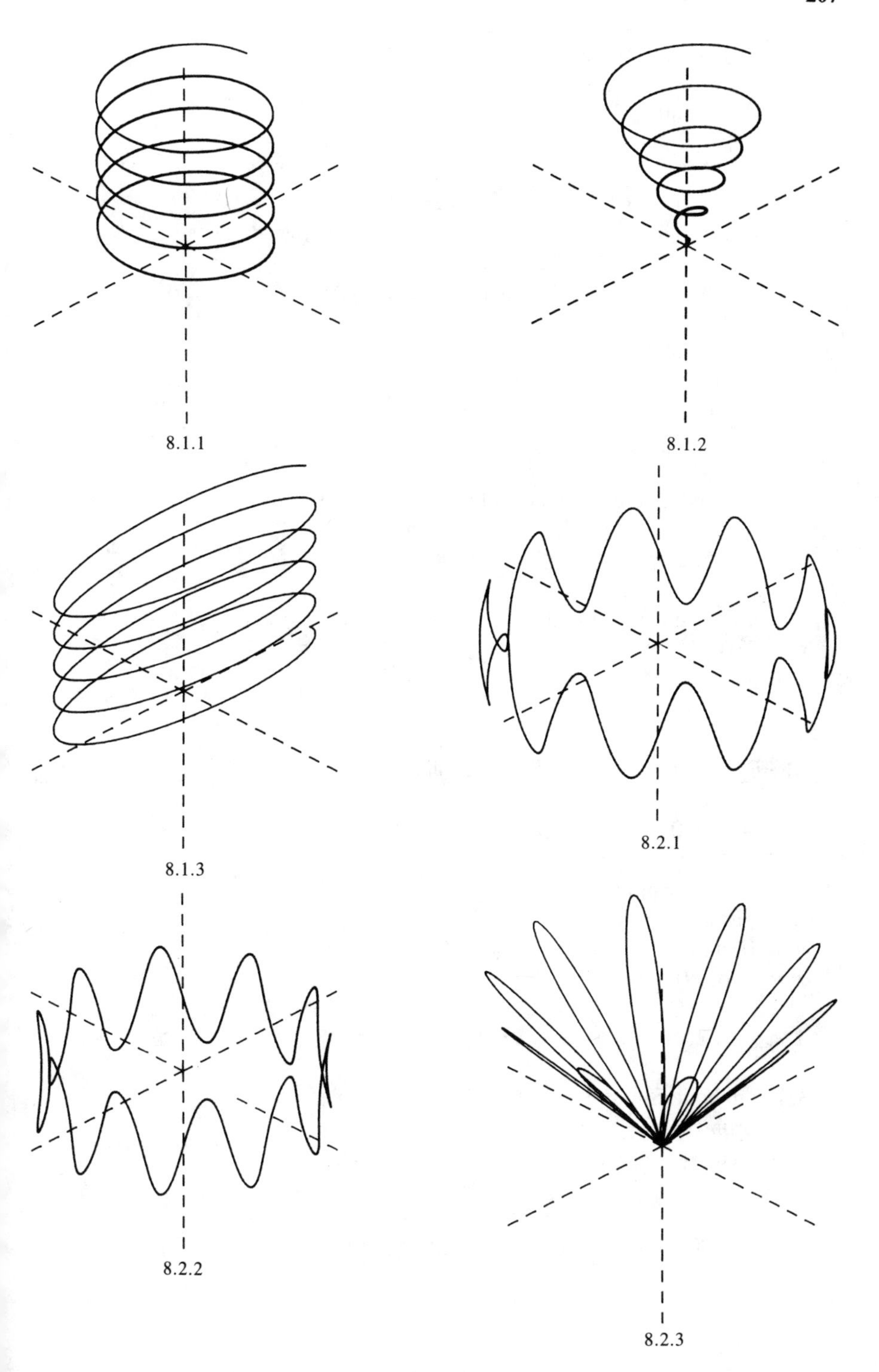
8.1.1
8.1.2
8.1.3
8.2.1
8.2.2
8.2.3

8.2.4. Rotating Sine Wave

$x = \sin(at)\cos(bt)$
$y = \sin(at)\sin(bt)$
$z = t/2\pi$

1. $a = 3.0$, $b = 1.00$, $-2\pi < t < 2\pi$; $\theta = 315$, $\phi = 60$
2. $a = 3.0$, $b = 0.25$, $-2\pi < t < 2\pi$; $\theta = 315$, $\phi = 60$

8.3. MISCELLANEOUS SPIRALS

8.3.1. Sici Spiral

$x = a \cdot Ci(t)$
$y = a \cdot Si(t)$
$z = t/c$

where Si and Ci are the Sine and Cosine integrals

1. $a = 0.5$, $c = 20.0$; $0 < t < 20$; $\theta = 300$, $\phi = 50$

8.3.2. Fresnel Integral Spiral

$x = C(t)$
$y = S(t)$
$z = t/c$

where S and C are the Fresnel integrals

1. $c = 5.0$; $0 < t < 5$; $\theta = 300$, $\phi = 50$

8.3.3. Toroidal Spiral

$x = [a \cdot \sin(ct) + b]\cos(t)$
$y = [a \cdot \sin(ct) + b]\sin(t)$
$z = a \cdot \cos(ct)$

1. $a = 0.2$, $b = 0.8$, $c = 20.0$; $0 < t < 2\pi$; $\theta = 315$, $\phi = 60$

8.3.4. Parabolic Spiral

$x = a \cdot \sin(t)$
$y = a \cdot \cos(t)$
$z = t^2/(2\pi c)^2$

1. $a = 0.5$, $c = 5.0$; $0 < t < 10\pi$; $\theta = 315$, $\phi = 60$

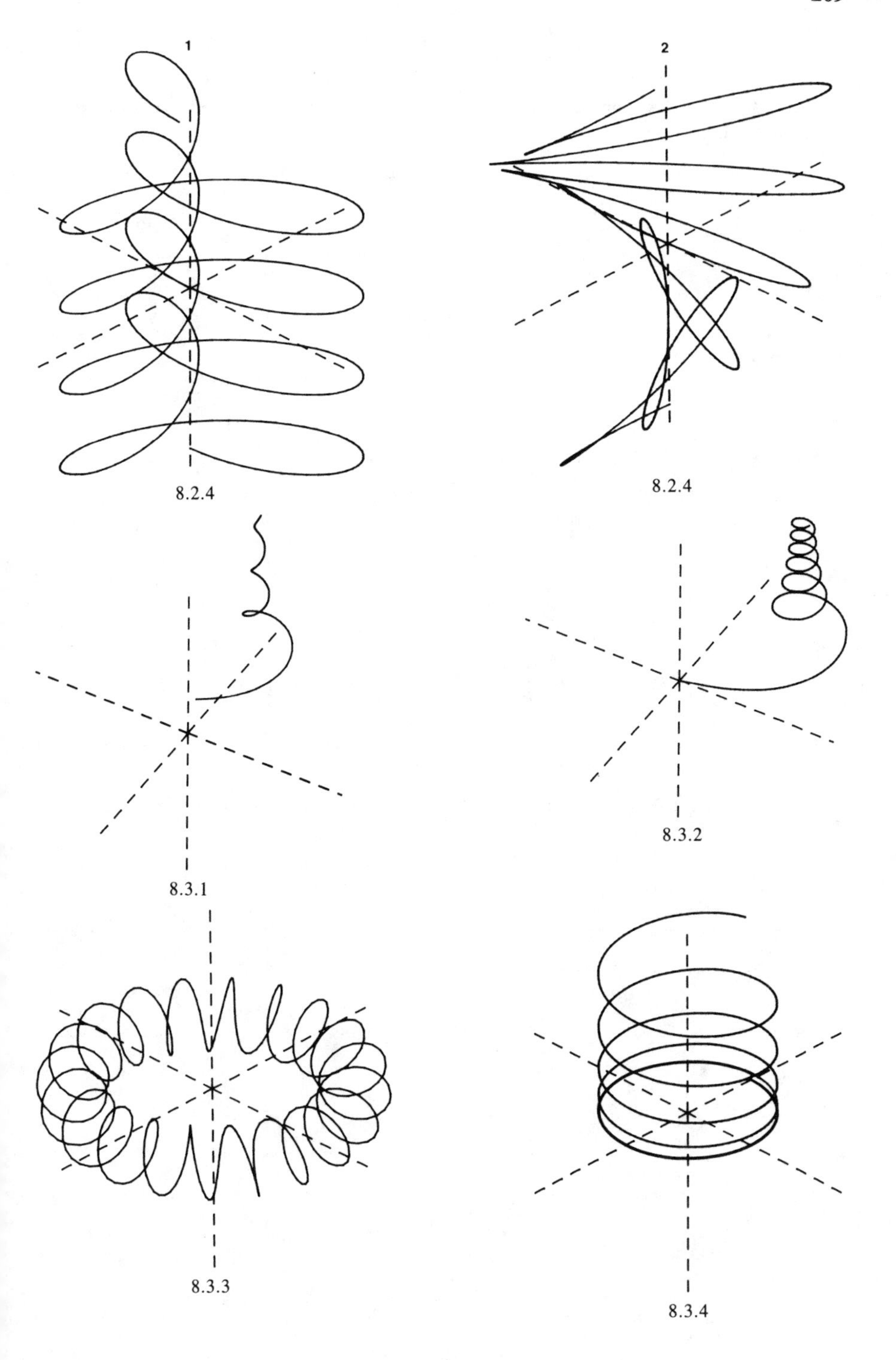

8.2.4

8.2.4

8.3.1

8.3.2

8.3.3

8.3.4

REFERENCE

1. **Foley, J. D. and VanDam, A.,** *Fundamentals of Interactive Computer Graphics,* Addison-Wesley, Reading, MA, 1983.

Chapter 9

ALGEBRAIC SURFACES

The following forms are plotted in the orthographic projection described in the text in Chapter 8. The plotted limits of the axes extend from -1.0 to $+1.0$ in all three dimensions as in Chapter 8 also. The plotted representation of surfaces on a two-dimensional projection plane has been given various treatments; here these conventions are used (except for the equations in Section 9.1):

1. Curves representing the intersection of planes of constant z with the surface are plotted at intervals (not necessarily regular) appropriate to each surface. (Curves at $z = \pm 1$ may or may not be plotted.)
2. The curves given by the intersection of the $y = 0$ and $x = 0$ planes with the surface are plotted wherever possible.
3. Any hidden portion of the lines representing the surface is plotted as a dotted, rather than a solid, line. "Hidden" is defined as simply wherever the normal to the surface forms an angle greater than 90° with the vector from the origin (0,0,0) to the viewpoint at infinity, defined by the angles (θ,ϕ). Note that this definition does not correctly define all "hidden" portions of a surface, unless it is a "convex" surface. However, for the relatively simple surfaces of this chapter, it is sufficient when the viewing angles (θ,ϕ) are carefully chosen.

9.1. FUNCTIONS WITH ax + by

9.1.1. $z = ax + by$ $ax + by - z = 0$
"Plane"

1. $a = 0.5$, $b = 0.5$; $\theta = 315$, $\phi = 70$
2. $a = 0.1$, $b = 0.3$; $\theta = 315$, $\phi = 70$

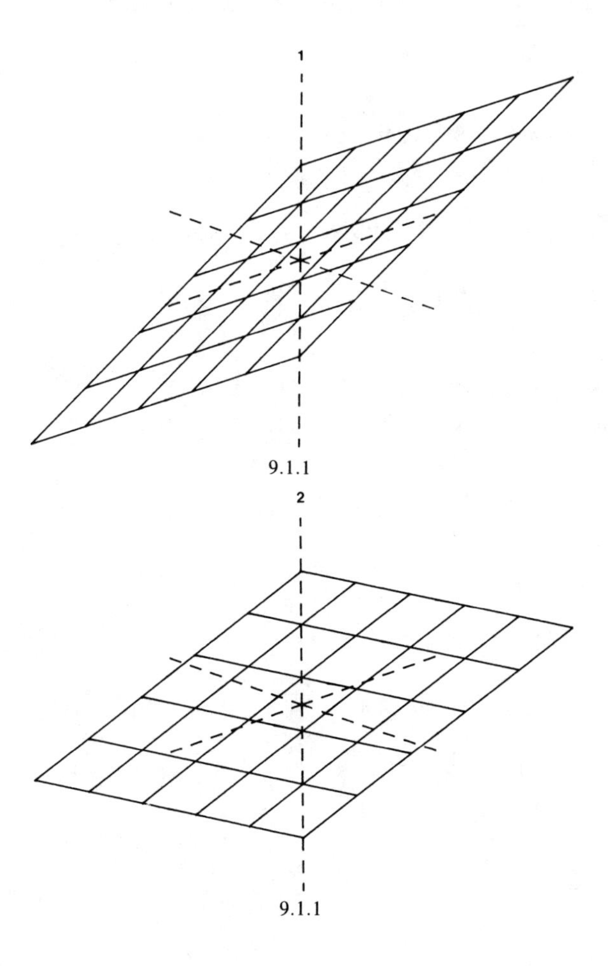

9.1.1

9.1.1

9.1.2. $z = 1/(ax + by)$ $axz + byz - 1 = 0$

1. $a = 5.0$, $b = 5.0$; $\theta = 240$, $\phi = 70$
2. $a = 2.0$, $b = 4.0$; $\theta = 240$, $\phi = 70$

9.2. FUNCTIONS WITH $x^2/a^2 \pm y^2/b^2$

9.2.1. $z = c(x^2/a^2 + y^2/b^2)$ $x^2/a^2 + y^2/b^2 - z/c = 0$
"Elliptic paraboloid"

1. $a = 0.5$, $b = 1.0$, $c = 1.0$; $\theta = 315$, $\phi = 60$
2. $a = 1.0$, $b = 1.0$, $c = 2.0$; $\theta = 315$, $\phi = 60$

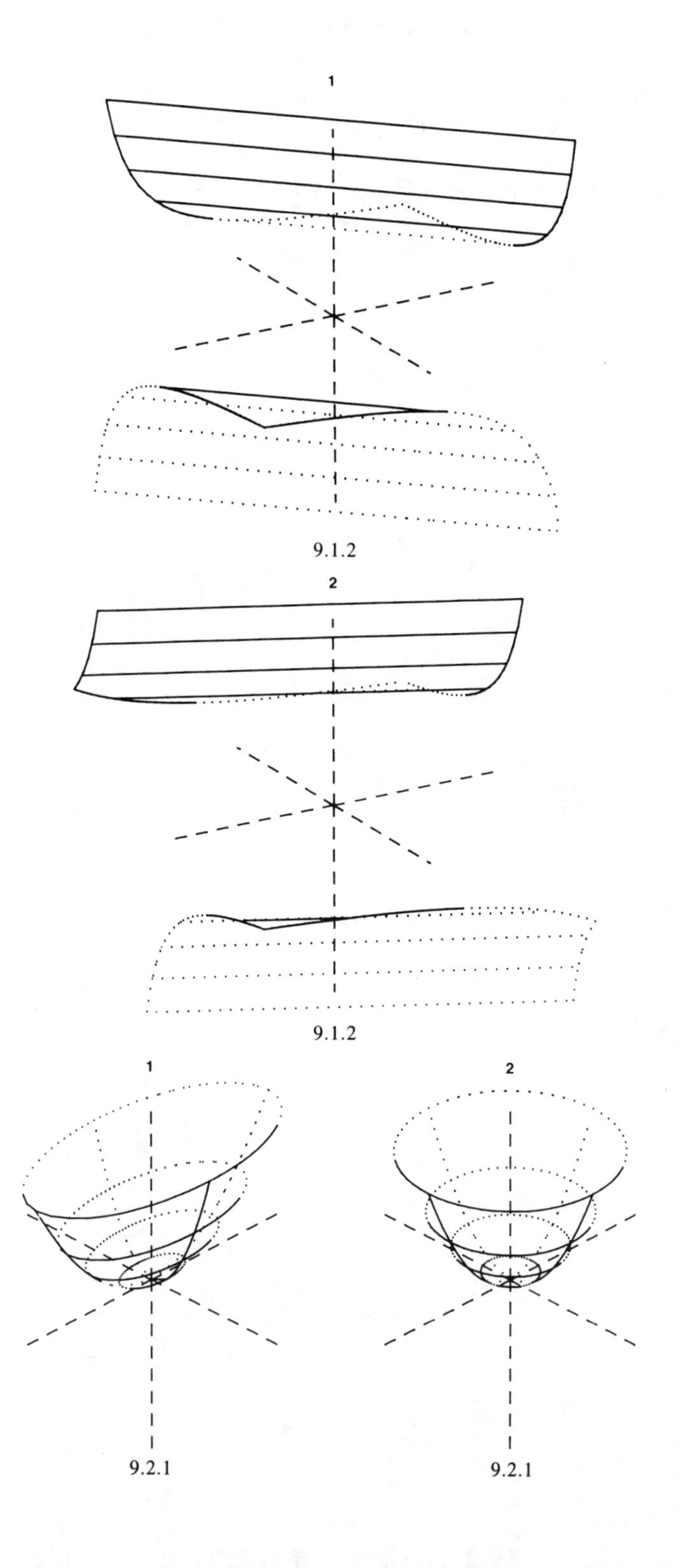

9.1.2

9.1.2

9.2.1

9.2.1

9.2.2. $z = c(x^2/a^2 - y^2/b^2)$ $x^2/a^2 - y^2/b^2 - z/c = 0$

"Hyperbolic paraboloid" (commonly called "saddle")

1. $a = 0.50$, $b = 0.5$, $c = 1.0$; $\theta = 300$, $\phi = 60$
2. $a = 0.75$, $b = 0.5$, $c = 1.0$; $\theta = 300$, $\phi = 60$

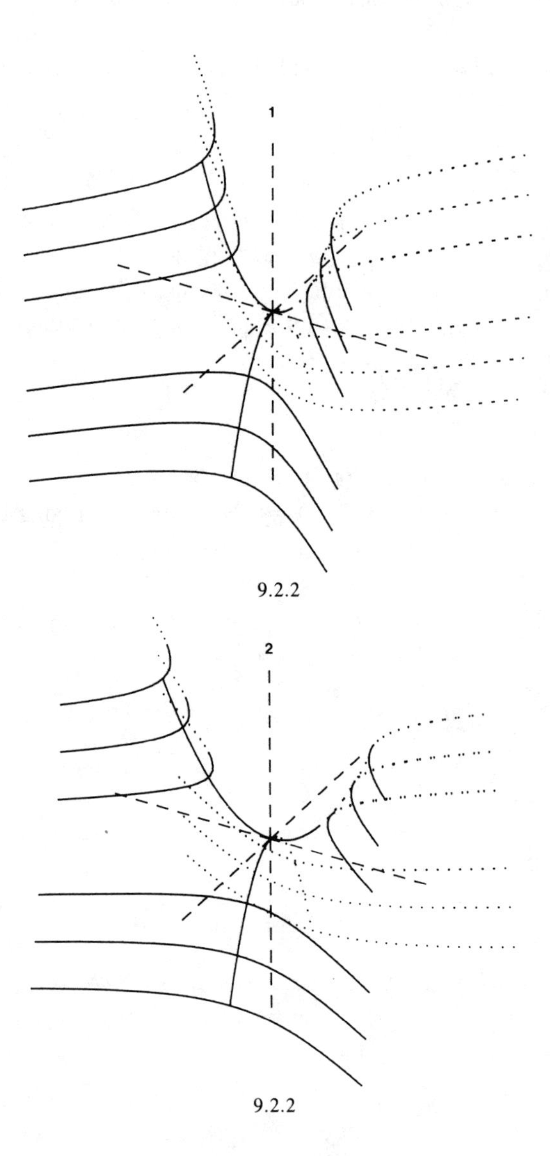

9.2.2

9.2.2

9.3. FUNCTIONS WITH $(x^2/a^2 + y^2/b^2 \pm c^2)^{1/2}$

9.3.1. $z = c(1 - x^2/a^2 - y^2/b^2)^{1/2}$ $\quad$ $x^2/a^2 + y^2/b^2 + z^2/c^2 - 1 = 0$
"Ellipsoid"

1. $a = 1.00$, $b = 1.00$, $c = 1.0$; $\theta = 300$, $\phi = 70$ (sphere)
2. $a = 0.50$, $b = 0.75$, $c = 1.0$; $\theta = 300$, $\phi = 70$
3. $a = 1.00$, $b = 1.00$, $c = 0.5$; $\theta = 300$, $\phi = 70$ (oblate spheroid)
4. $a = 0.50$, $b = 0.50$, $c = 1.0$; $\theta = 300$, $\phi = 70$ (prolate spheroid)

9.3.2. $z = c(x^2/a^2 + y^2/b^2)^{1/2}$ $\quad$ $x^2/a^2 + y^2/b^2 - z^2/c^2 = 0$
"Elliptic cone"

1. $a = 0.5$, $b = 1.0$, $c = 1.00$; $\theta = 315$, $\phi = 70$
2. $a = 1.0$, $b = 1.0$, $c = 0.75$; $\theta = 315$, $\phi = 70$ (circular cone)

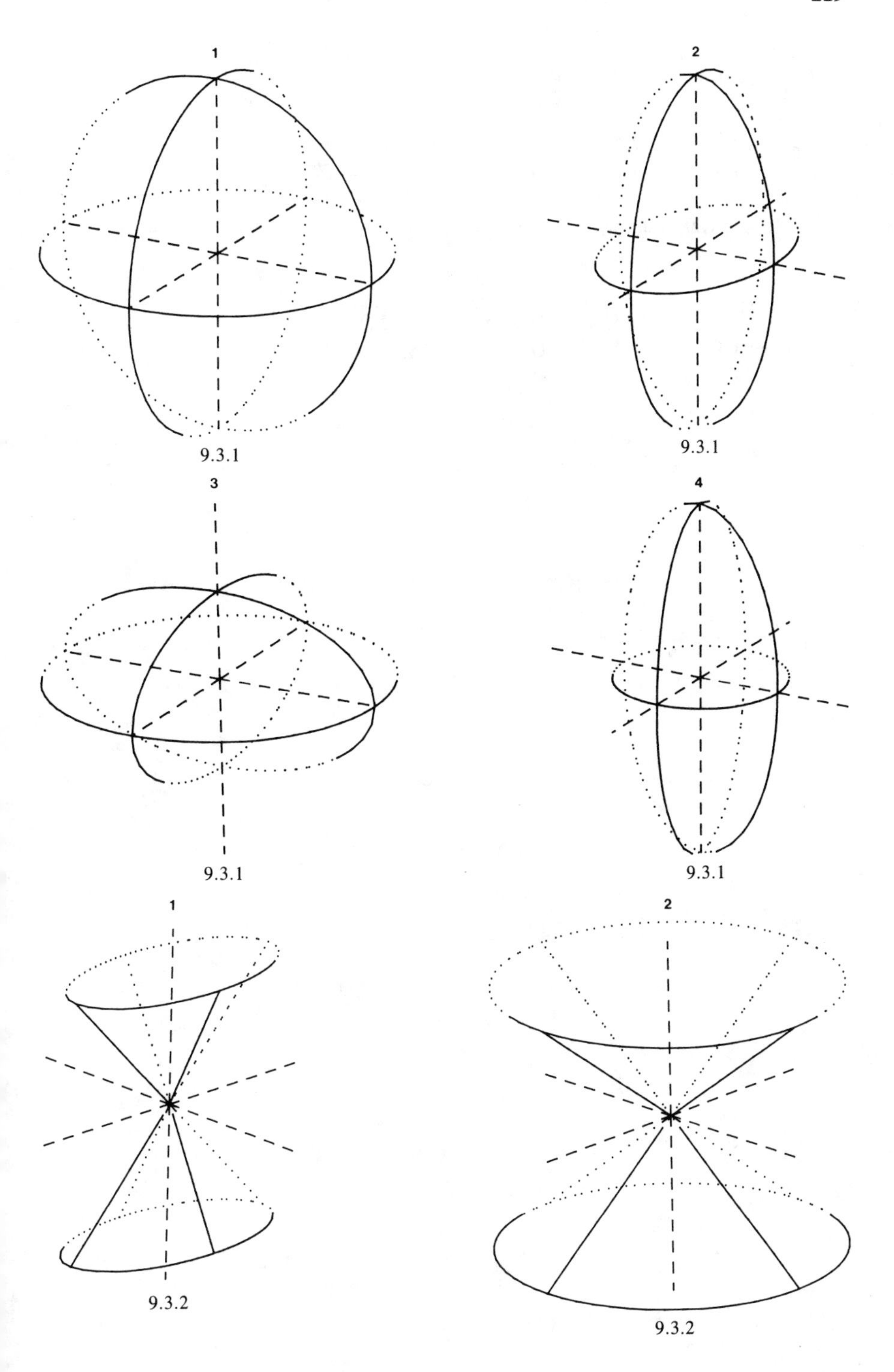

9.3.1

9.3.1

9.3.1

9.3.1

9.3.2

9.3.2

9.3.3. $z = c(x^2/a^2 + y^2/b^2 - 1)^{1/2}$ $x^2/a^2 + y^2/b^2 - z^2/c^2 - 1 = 0$
"Hyperboloid of one sheet"

1. $a = 0.5$, $b = 0.5$, $c = 1.0$; $\theta = 315$, $\phi = 70$
2. $a = 0.4$, $b = 0.2$, $c = 0.5$; $\theta = 315$, $\phi = 70$

9.3.4. $z = c(x^2/a^2 + y^2/b^2 + 1)^{1/2}$ $x^2/a^2 + y^2/b^2 - z^2/c^2 + 1 = 0$
"Hyperboloid of two sheets"

1. $a = 0.3$, $b = 0.3$, $c = 0.3$; $\theta = 315$, $\phi = 70$
2. $a = 0.1$, $b = 0.2$, $c = 0.2$; $\theta = 315$, $\phi = 70$

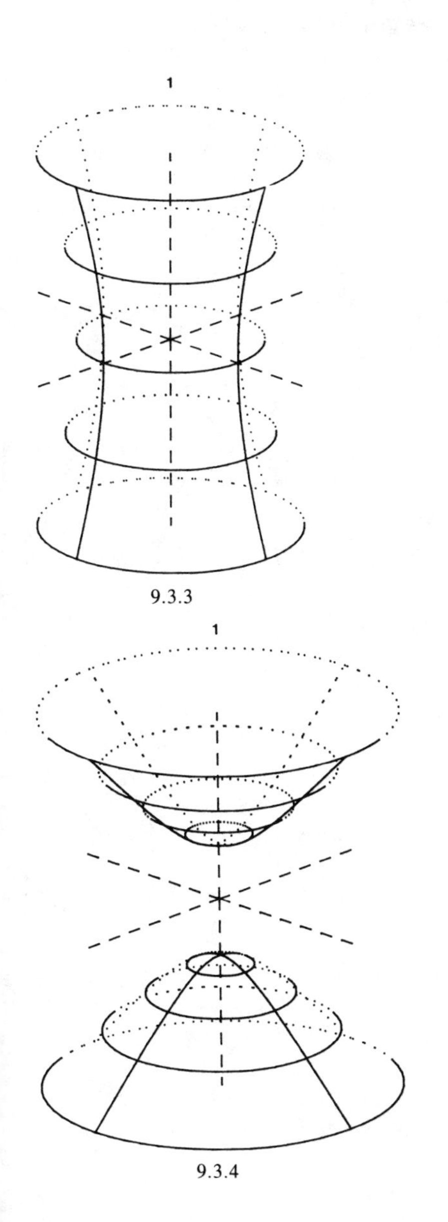

9.3.3

9.3.4

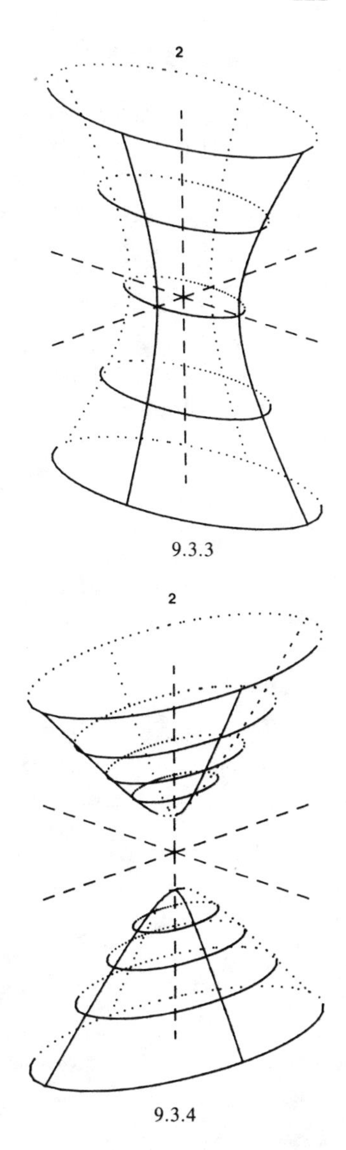

9.3.3

9.3.4

9.4. FUNCTIONS WITH $x^3/a^3 \pm y^3/b^3$

9.4.1. $z = c(x^3/a^3 + y^3/b^3)$
1. $a = 1.0$, $b = 1.0$, $c = 2.0$; $\theta = 190$, $\phi = 70$

9.4.2. $z = c(x^3/a^3 - y^3/b^3)$
1. $a = 1.0$, $b = 1.0$, $c = 1.0$; $\theta = 190$, $\phi = 70$

9.5. FUNCTIONS WITH $x^4/a^4 \pm y^4/b^4$

9.5.1. $z = c(x^4/a^4 + y^4/b^4)$
1. $a = 1.0$, $b = 1.0$, $c = 3.0$; $\theta = 310$, $\phi = 70$

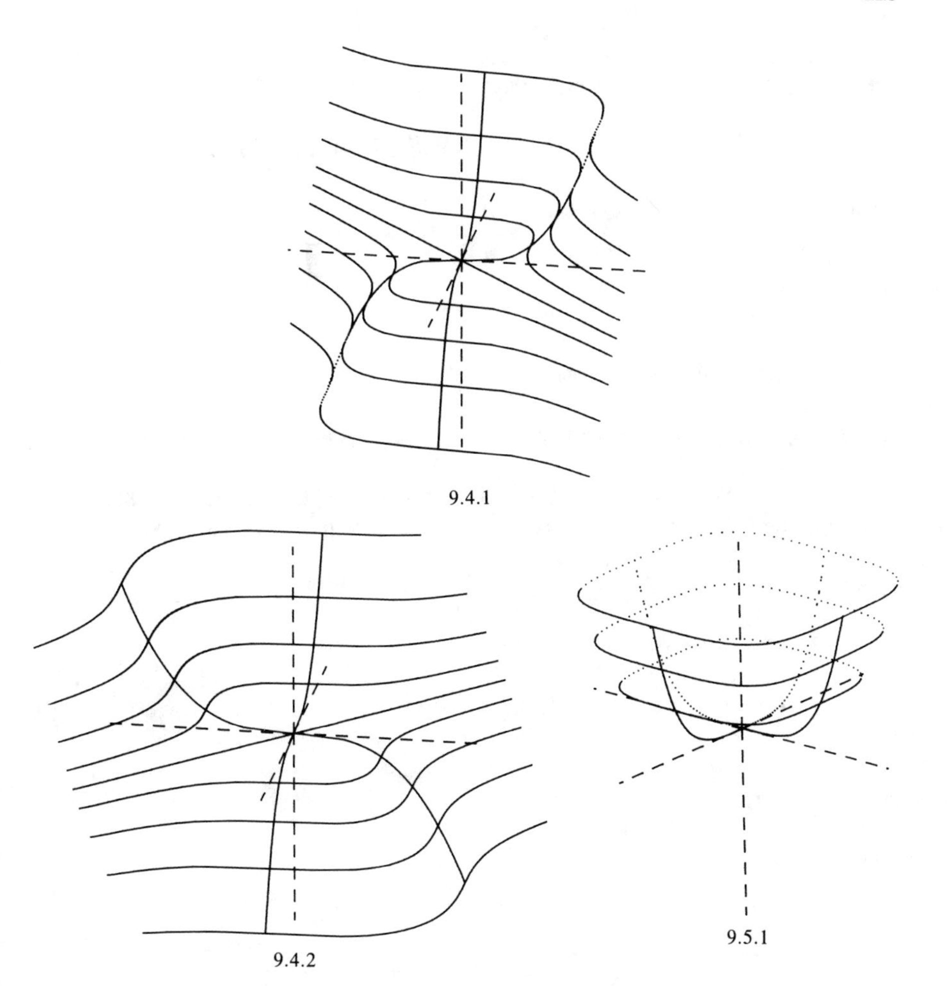

9.4.1

9.4.2

9.5.1

9.5.2. $z = c(x^4/a^4 - y^4/b^4)$
1. $a = 0.5$, $b = 0.5$, $c = 1.0$; $\theta = 300$, $\phi = 60$

9.6. MISCELLANEOUS FUNCTIONS

9.6.1. $z = \{r^2 - [(x^2 + y^2)^{1/2} - b]^2\}^{1/2}$
"Torus"

1. $a = 0.2$, $b = 0.8$; $\theta = 300$, $\phi = 60$

9.6.2. $z = c[\arctan(y/x) + n\pi]$
"Right helicoid"
Parametrically, $x = u \cdot \cos(v)$; $y = u \cdot \sin(v)$; $z = cv$

1. $c = 1/(2\pi)$, $-2\pi < v < 2\pi$, $-0.5 < u < 0.5$; $\theta = 315$, $\phi = 80$
2. $c = 1/(2\pi)$, $-2\pi < v < 2\pi$, $0.3 < u < 0.6$; $\theta = 315$, $\phi = 80$

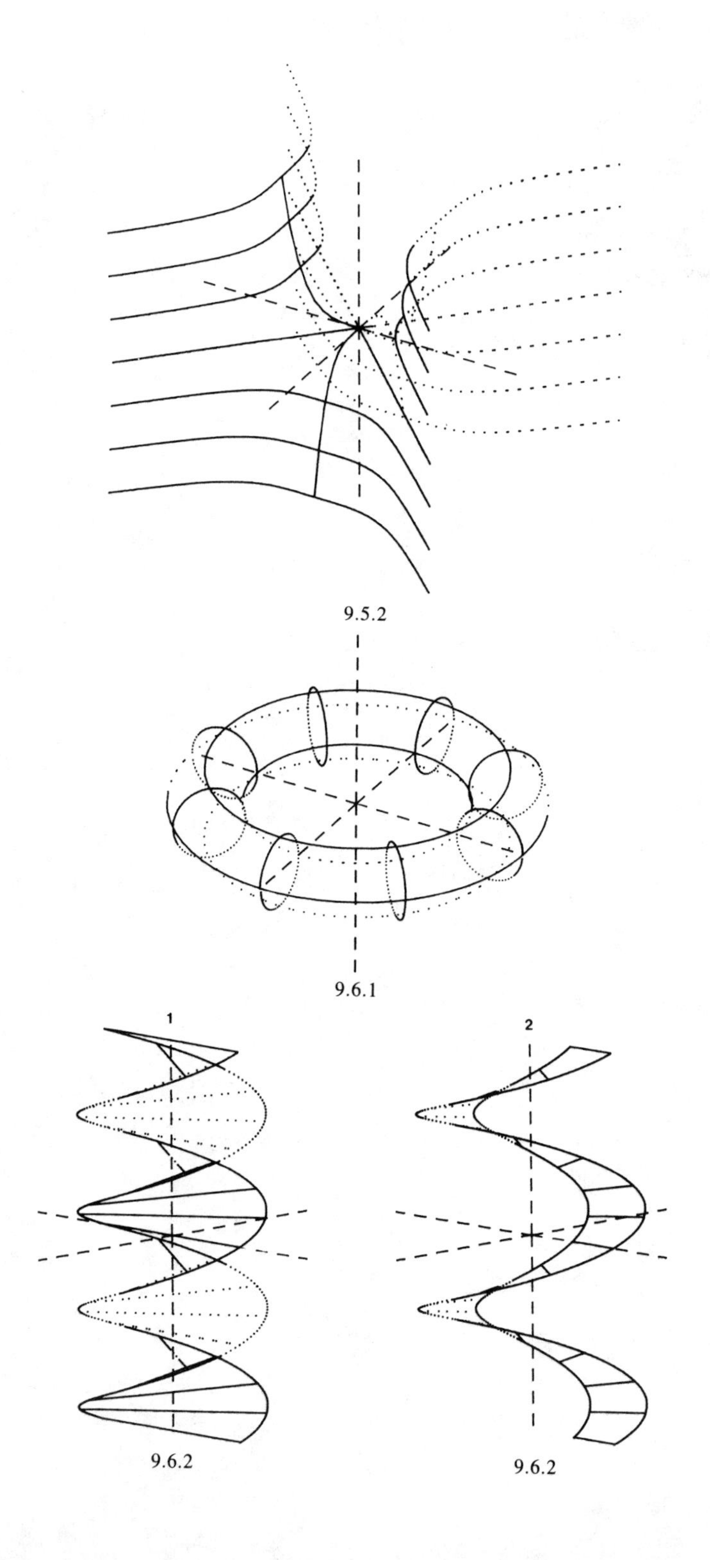

9.5.2

9.6.1

9.6.2

9.6.2

9.6.3. $z = c(x^3 - 3xy^2)$
"Monkey saddle"

1. $c = 1.0$; $\theta = 330$, $\phi = 60$

9.6.4. $z = cx^2y^2$
"Crossed trough"

1. $c = 4.0$; $\theta = 300$, $\phi = 60$

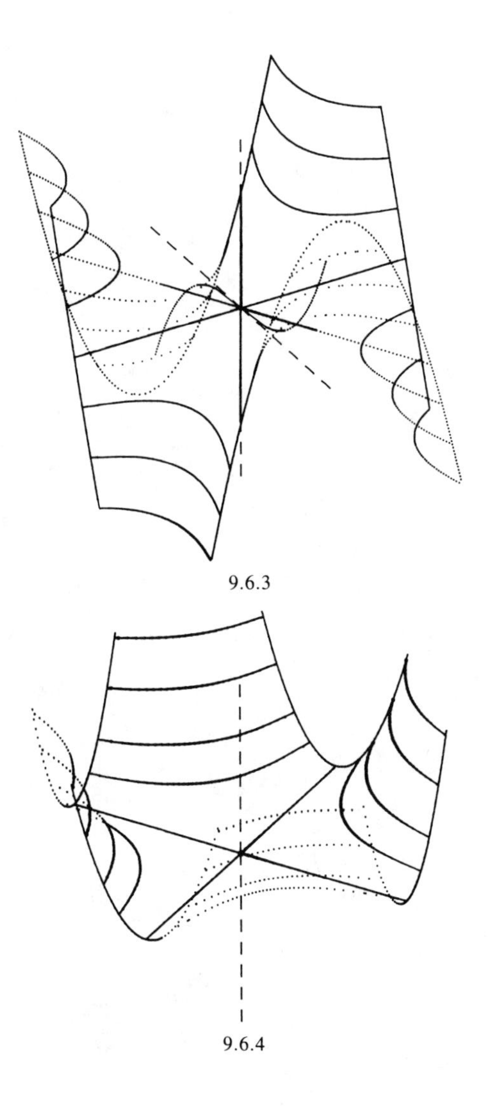
9.6.3
9.6.4

Chapter 10

TRANSCENDENTAL SURFACES

The orthographic projection (Chapter 8 text) is again used here, along with the axes limits of ± 1. The treatment of surfaces for this chapter (except for those in Section 10.5) varies from that used in Chapter 9. Here a more complete representation of the surfaces is attained by a ''hidden-line'' algorithm which is applied to lines formed by the intersection of the surface with planes of constant x value. This algorithm (found in, for example, Hearn and Baker[1]) eliminates any part of the line which is hidden by a portion of the surface between the line segment and the viewpoint at infinity. The steps in generating the surfaces are as follows:

1. Generate a line representing the surface $z = f(x,y)$ at a constant x. Clip this line at $z = \pm 1$.
2. Project the line to the viewing plane by the equations in the text in Section 8.
3. Apply the hidden-line algorithm in plotting the line.
4. Increment x and repeat (1) to (3).

10.1. TRIGONOMETRIC FUNCTIONS

10.1.1. $z = c \cdot \sin[2\pi a(x^2 + y^2)^{1/2}]$
1. $a = 3.0$, $c = 0.25$; $\theta = 325$, $\phi = 50$; $(x^2 + y^2)^{1/2} < 1$

10.1.2. $z = c \cdot \cos[2\pi a(x^2 + y^2)^{1/2}]$
1. $a = 3.0$, $c = 0.25$; $\theta = 325$, $\phi = 50$; $(x^2 + y^2)^{1/2} < 1$

10.1.3. $z = c \cdot \sin(2\pi axy)$
1. $a = 3.0$, $c = 0.25$; $\theta = 325$, $\phi = 50$

10.1.4. $z = c \cdot \cos(2\pi axy)$
1. $a = 3.0$, $c = 0.25$; $\theta = 325$, $\phi = 50$

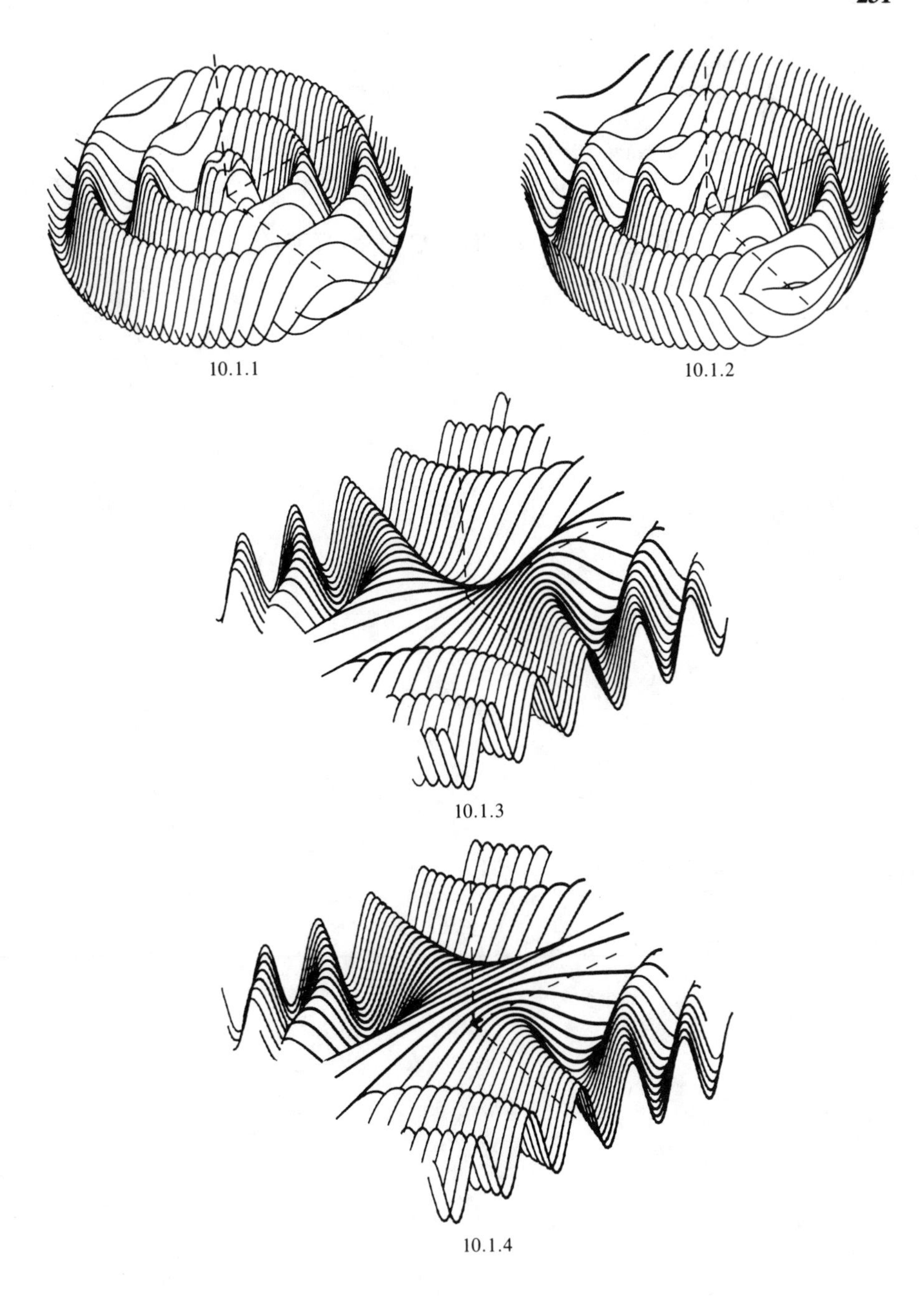

10.1.1

10.1.2

10.1.3

10.1.4

10.1.5. $z = c \cdot \sin(2\pi ax)\sin(2\pi by)$
1. $a = 2.0$, $b = 1.0$, $c = 0.25$; $\theta = 325$, $\phi = 50$

10.1.6. $z = c \cdot \cos(2\pi ax)\cos(2\pi by)$
1. $a = 2.0$, $b = 1.0$, $c = 0.25$; $\theta = 325$, $\phi = 50$

10.2. LOGARITHMIC FUNCTIONS

10.2.1. $z = c \cdot \ln(ax^2 + by^2)$
1. $a = 1.0$, $b = 2.0$, $c = 0.2$; $\theta = 330$, $\phi = 50$

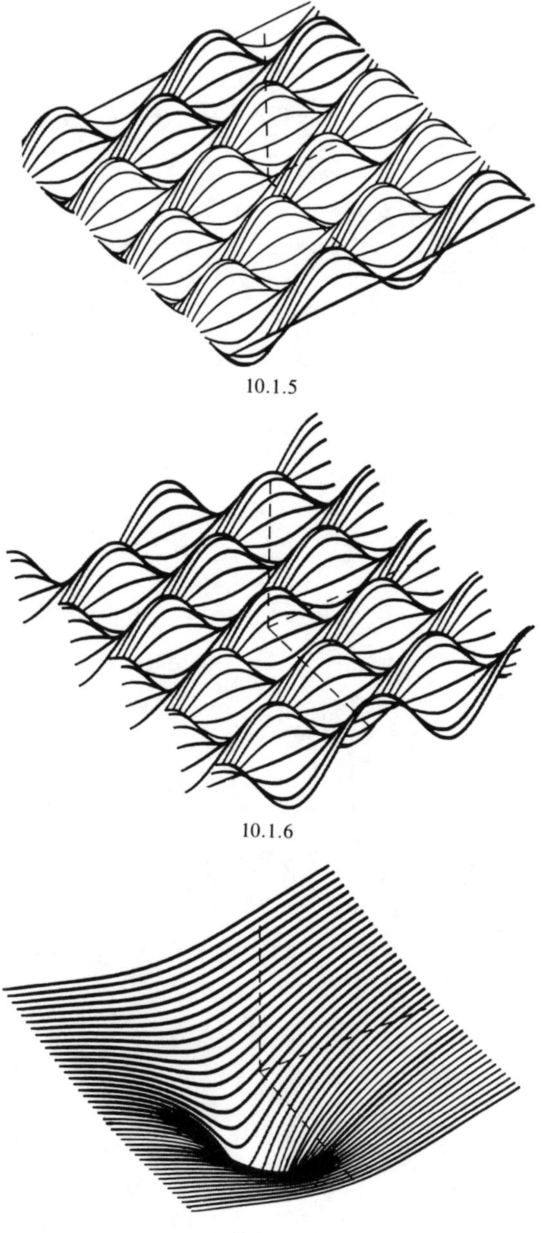

10.1.5

10.1.6

10.2.1

10.2.2. $z = c \cdot \ln(|xy|)$
1. $c = 0.2$; $\theta = 330$, $\phi = 50$

10.3. EXPONENTIAL FUNCTIONS

10.3.1. $z = c \cdot \exp(ax + by)$
1. $a = 2.0$, $b = 2.0$, $c = 0.25$; $\theta = 315$, $\phi = 60$
2. $a = 2.0$, $b = -2.0$, $c = 0.25$; $\theta = 30$, $\phi = 50$

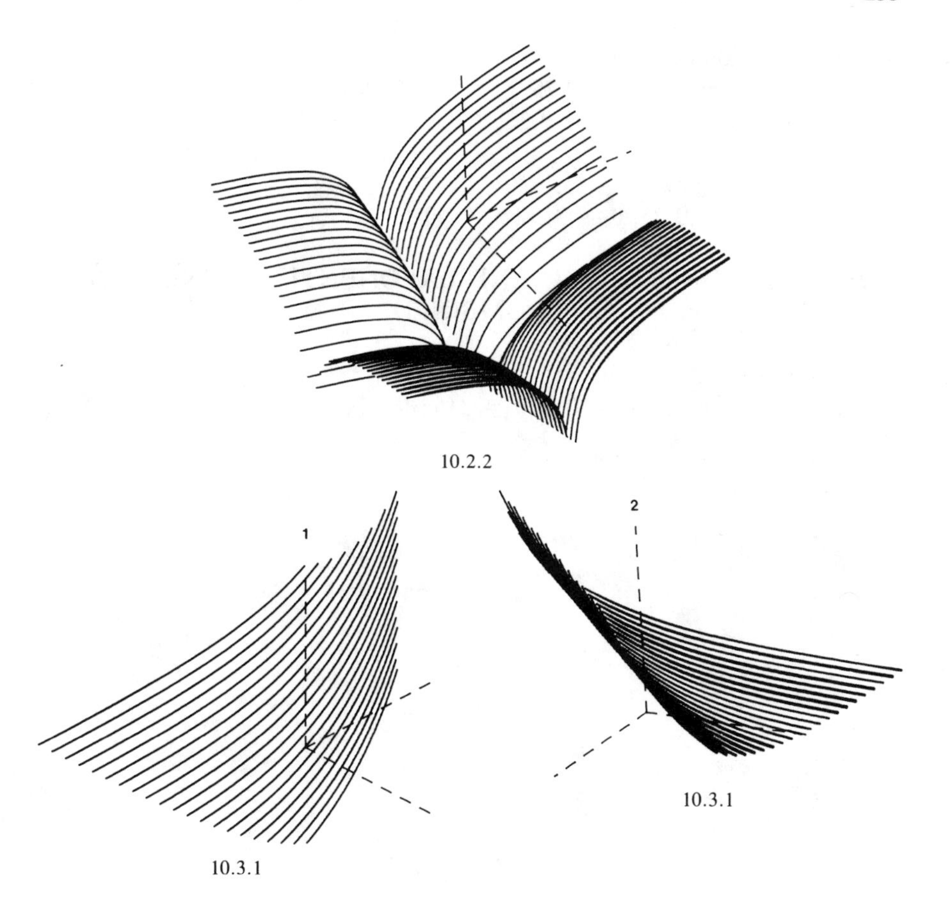

10.2.2

10.3.1

10.3.1

10.3.2. $z = c \cdot \exp(ax^2 + by^2)$

1. $a = 1.0$, $b = 2.0$, $c = 0.25$; $\theta = 315$, $\phi = 60$
2. $a = 1.0$, $b = -2.0$, $c = 0.30$; $\theta = 315$, $\phi = 60$

10.3.3. $z = c \cdot \exp(axy)$

1. $a = 3.0$, $c = 0.25$; $\theta = 315$, $\phi = 60$

10.4. TRIGONOMETRIC AND EXPONENTIAL FUNCTIONS COMBINED

10.4.1. $z = c \cdot \cos(2\pi ar)e^{-br}$

1. $a = 3.0$, $b = 2.0$, $c = 0.5$; $\theta = 315$, $\phi = 40$; $(x^2 + y^2)^{1/2} < 1$

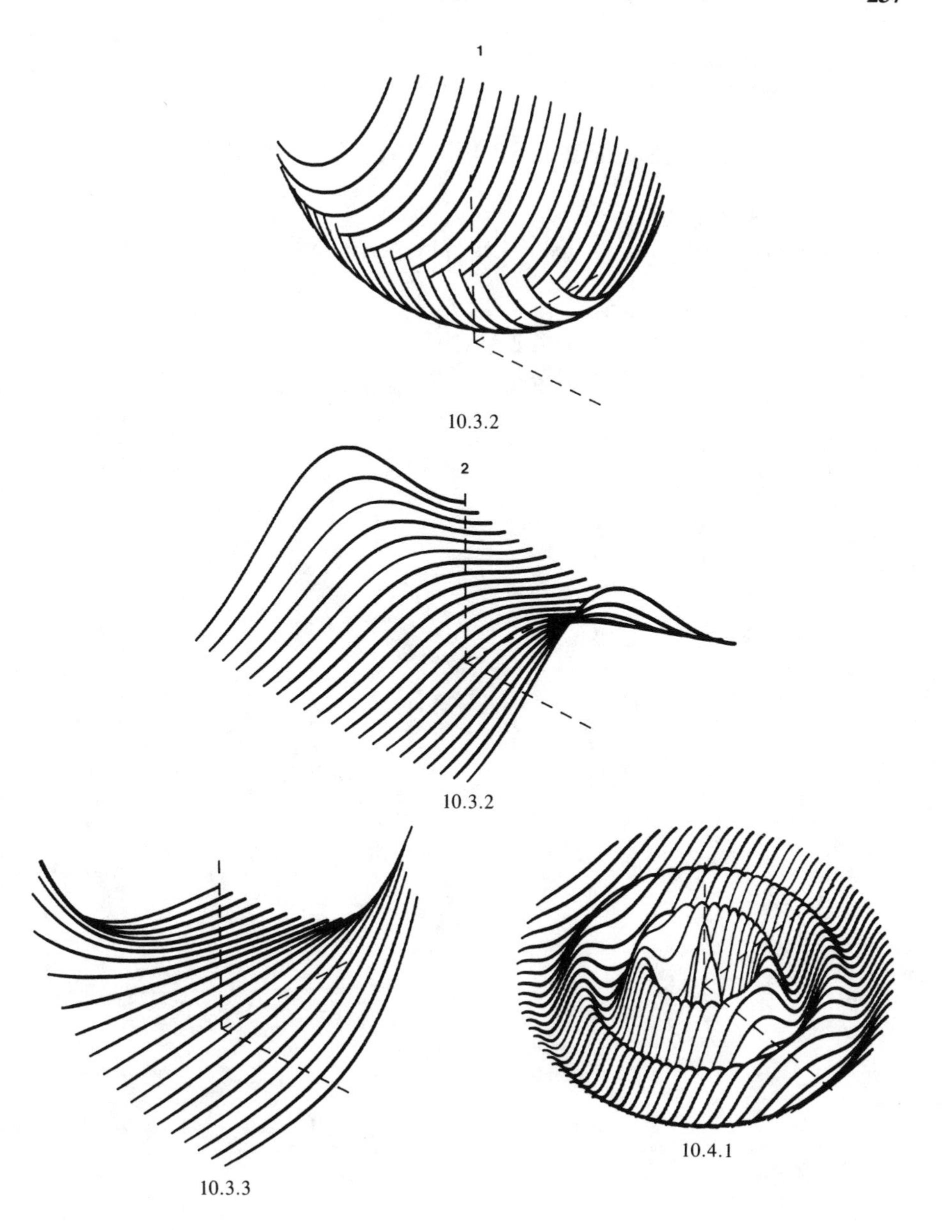

10.3.2

10.3.2

10.3.3

10.4.1

10.4.2. $z = c \cdot \sin(2\pi ar)e^{-br}$

1. $a = 3.0$, $b = 2.0$, $c = 0.5$; $\theta = 315$, $\phi = 40$; $(x^2 + y^2)^{1/2} < 1$

10.5. SURFACE SPHERICAL HARMONICS

10.5.1. $r = 1 + c \cdot P_n^0(\cos\phi)$

"Zonal harmonics" — P_n^0 is the Legendre polynomial

1. $n = 1$, $c = 0.20$; $\theta = 310$, $\phi = 70$
2. $n = 2$, $c = 0.10$; $\theta = 310$, $\phi = 70$
3. $n = 3$, $c = 0.20$; $\theta = 310$, $\phi = 70$

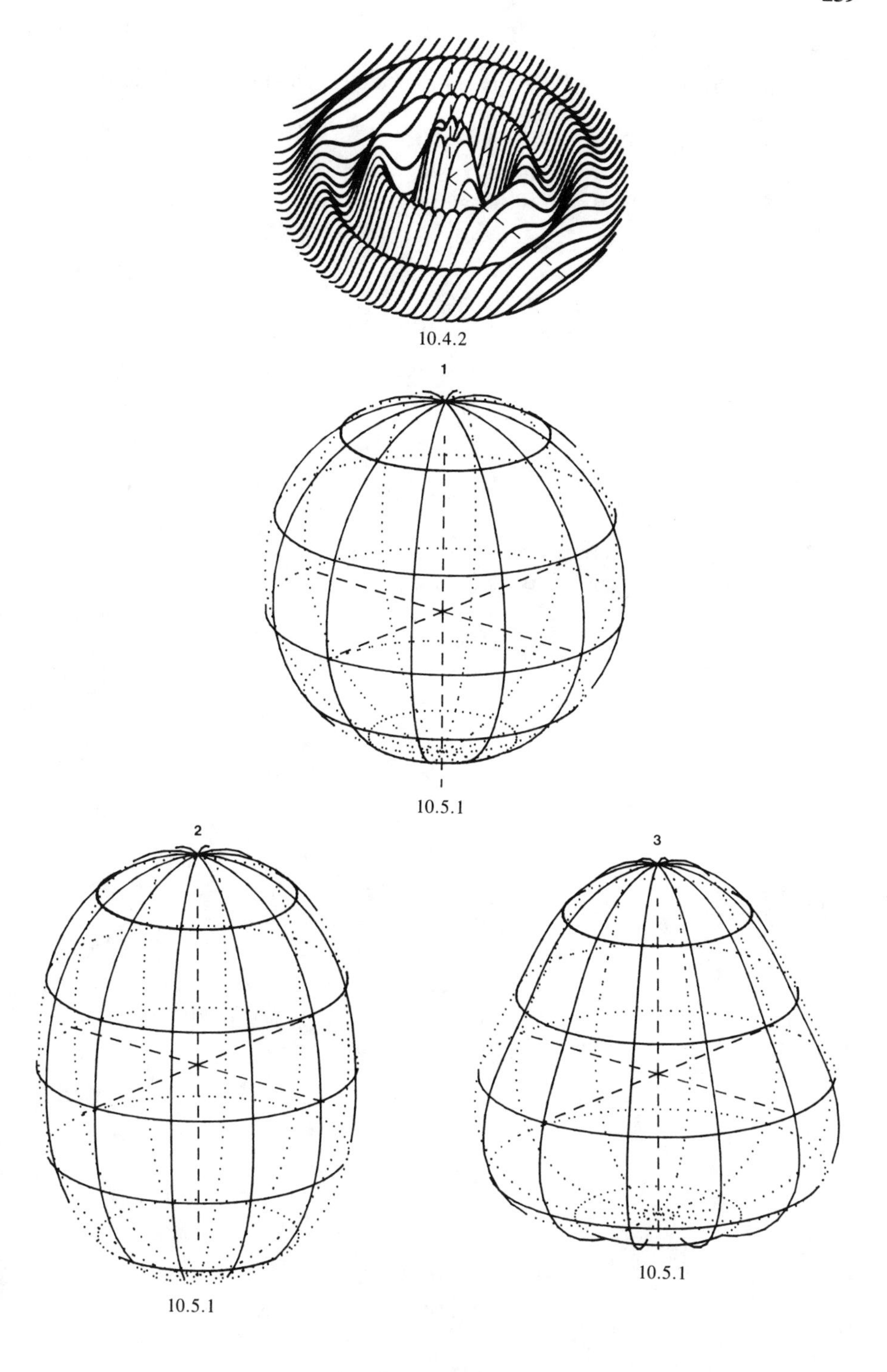

10.4.2

10.5.1

10.5.1

10.5.1

10.5.2. $r = 1 + c \cdot P_n^n(\cos\phi) \cdot \cos(n\theta)$

"Sectoral harmonics" — P_n^n is the associated Legendre function of the first kind

1. $n = 1$, $c = 0.20$; $\theta = 310$, $\phi = 70$
2. $n = 2$, $c = 0.10$; $\theta = 310$, $\phi = 70$
3. $n = 3$, $c = 0.02$; $\theta = 310$, $\phi = 70$

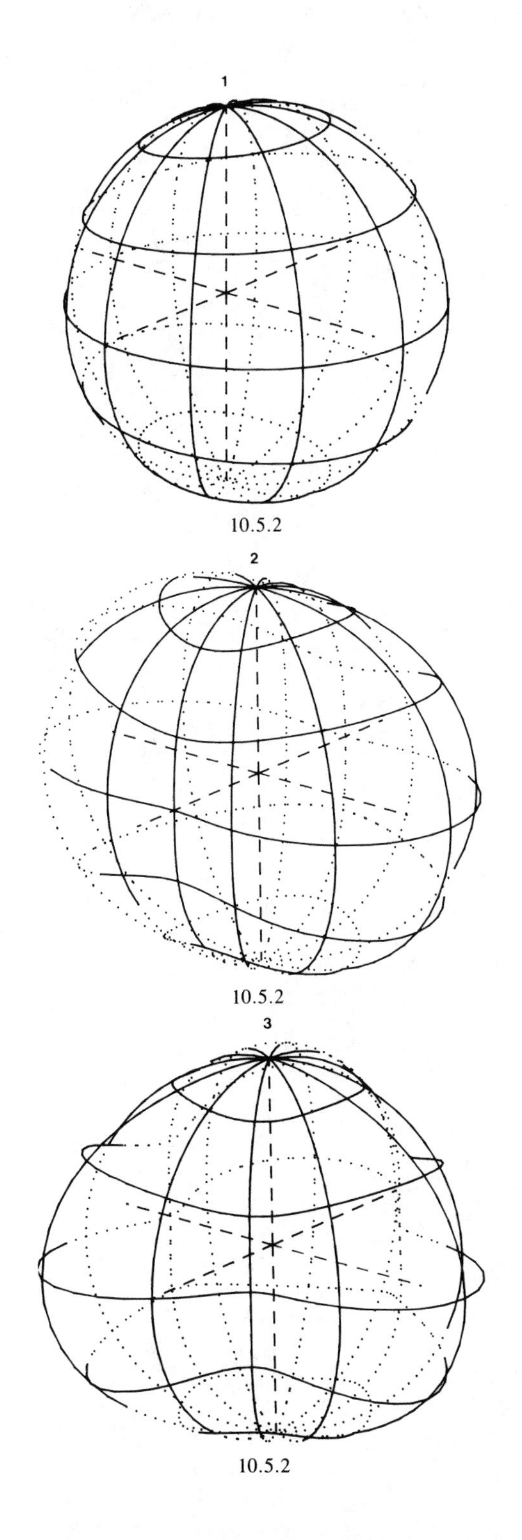

10.5.2

10.5.2

10.5.2

10.5.3. $r = 1 + c \cdot P_n^m(\cos\phi) \cdot \cos(m\theta)$

"Tesseral harmonics" — P_n^m is the associated Legendre function of the first kind

1. $n = 2$, $m = 1$, $c = 0.15$; $\theta = 310$, $\phi = 70$
2. $n = 3$, $m = 1$, $c = 0.10$; $\theta = 310$, $\phi = 70$
3. $n = 3$, $m = 2$, $c = 0.05$; $\theta = 310$, $\phi = 70$

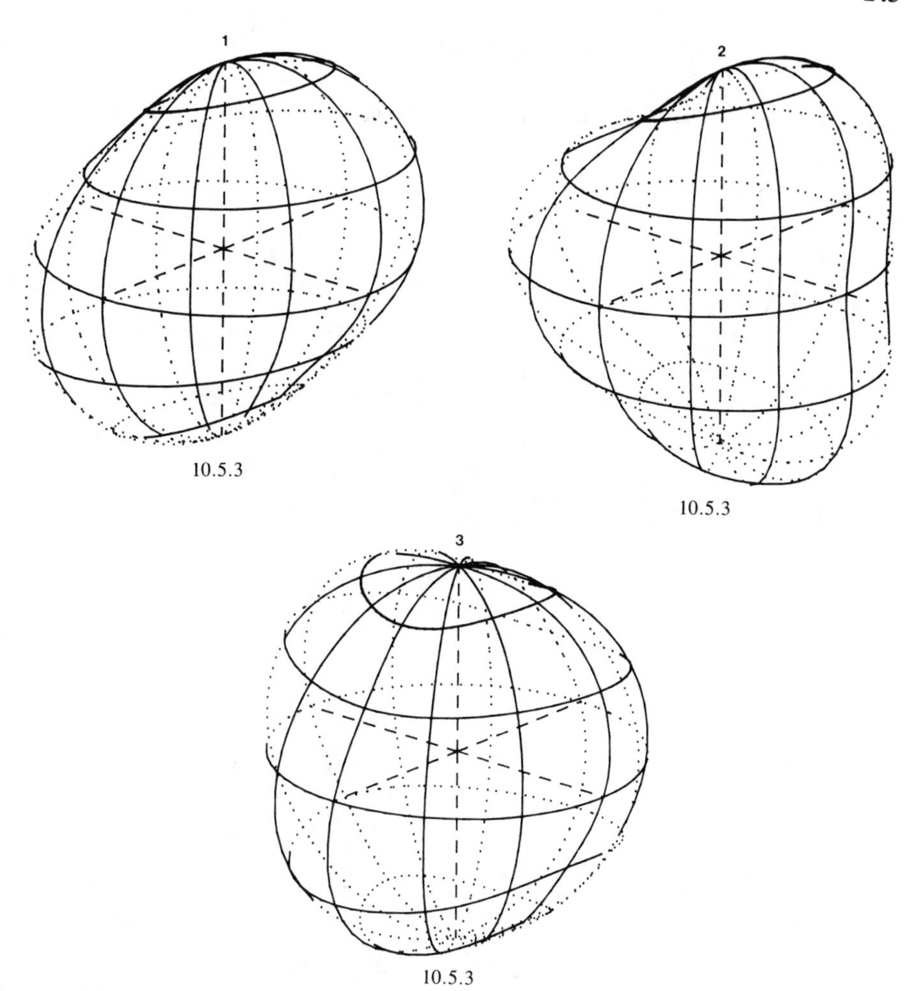

10.5.3

10.5.3

10.5.3

REFERENCE

1. **Hearn, D. and Baker, M. P.,** *Computer Graphics,* Prentice-Hall, Englewood Cliffs, NJ, 1986.

Chapter 11

NONDIFFERENTIABLE AND DISCONTINUOUS FUNCTIONS

In the equations of this chapter, the symbol H is used for the unit step function and the symbol δ for the unit impulse function. The function δ is defined only over an infinitesimal interval of x such that its integral over the infinitesimal interval is unity. This requires δ to have an infinite amplitude, and the amplitude is truncated here at $y = 1$ for purposes of illustration. The function H is defined such that H(a) is zero for $x < a$ and $H(a) = 1$ for $x > a$. Therefore, H(a) is the integral of $\delta(a)$.

11.1. FUNCTIONS WITH A FINITE NUMBER OF DISCONTINUITIES

11.1.1. $y = \delta(x - a)$
"Delta function"

1. $a = 0.5$

11.1.2. $y = \delta'(x - a)$
"Doublet function"

1. $a = 0.5$

11.1.3. $y = c[H(x - a)]$
"Step function"

1. $a = 0.5$, $c = 0.5$

11.1.4. $y = c[H(x - a) - H(x - b)]$
"Boxcar function"

1. $a = 0.25$, $b = 0.75$, $c = 0.50$

11.1.5. $y = c[H(x - a) - 2H(x - b) + H(x - 2b + a)]$
"Double boxcar function"

1. $a = 0.25$, $b = 0.50$, $c = 0.50$

11.2. FUNCTIONS WITH AN INFINITE NUMBER OF DISCONTINUITIES

11.2.1. $y = c \sum_{n=0}^{\infty} H(x - na)$

1. $a = 0.2$, $c = 0.1$

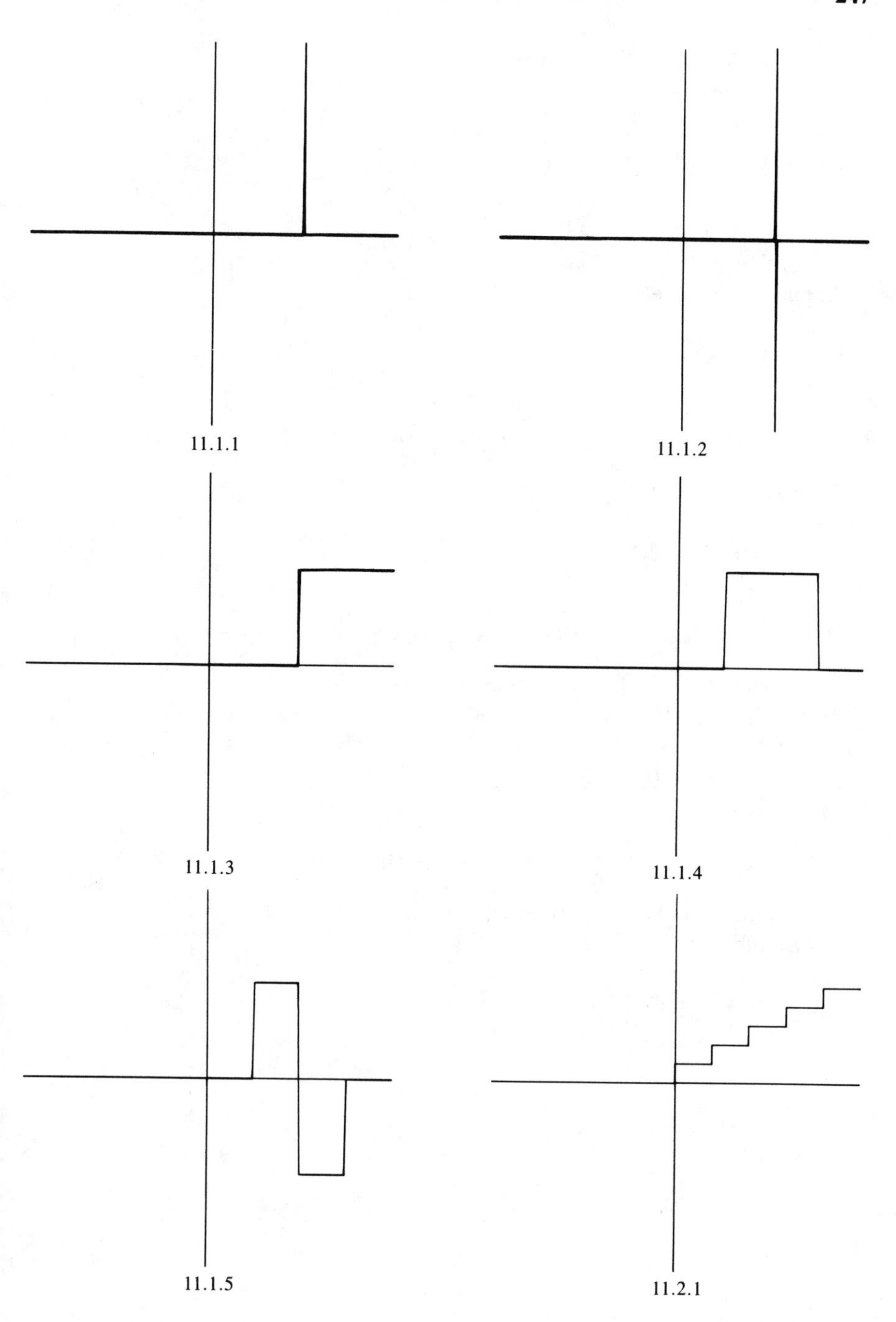
11.1.1
11.1.2
11.1.3
11.1.4
11.1.5
11.2.1

11.2.2. $y = \sum_{n=1}^{\infty} c^n H(x - na)$

1. $a = 0.2$, $c = 0.5$
2. $a = 0.2$, $c = -0.5$

11.2.3. $y = c[-1 + 2 \sum_{n=-\infty}^{\infty} (-1)^n H(x - na)]$

"Square sine wave"

1. $a = 0.1$, $c = 0.5$

11.2.4. $y = c\{-1 + 2 \sum_{n=-\infty}^{\infty} (-1)^n H[x - (n - 1/2) a]\}$

"Square cosine wave"

1. $a = 0.1$, $c = 0.5$

11.2.5. $y = c((2x/a) \sum_{n=-\infty}^{\infty} \{H[x - na] - H[x - (n + 1)a]\} - 2 \sum_{n=-\infty}^{\infty} n\{H[x - na] - H[x - (n + 1)a]\} - 1)$

"Sawtooth wave"

1. $a = 0.2$, $c = 0.5$

11.2.6. $y = c((-2x/a) \sum_{n=-\infty}^{\infty} \{H[x - na] - H[x - (n + 1) a]\} + 2 \sum_{n=-\infty}^{\infty} (n + 1)\{H[x - na] - H[x - (n + 1)a]\} - 1)$

"Sawtooth wave"

1. $a = 0.2$, $c = 0.5$

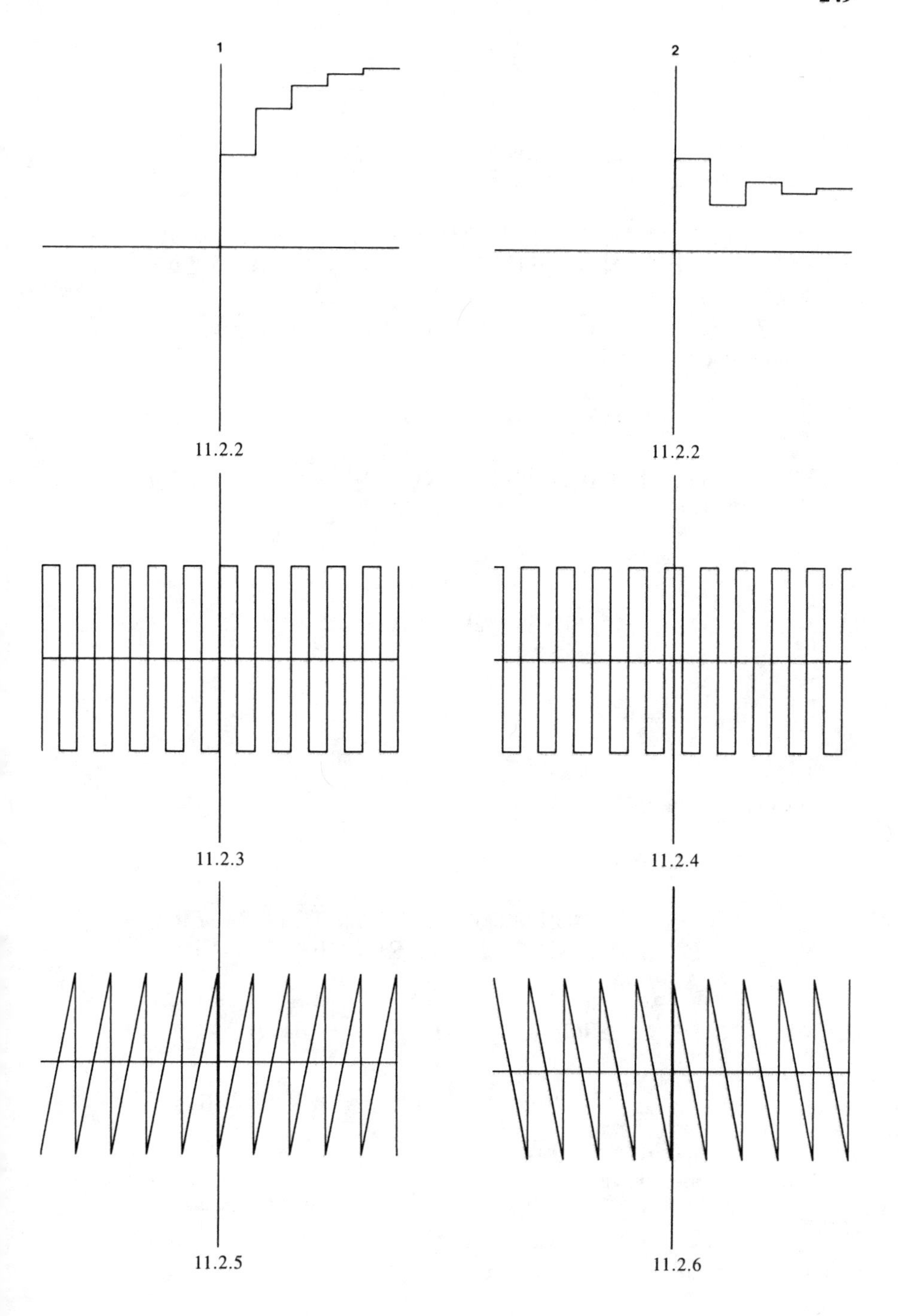
1
2
11.2.2
11.2.2
11.2.3
11.2.4
11.2.5
11.2.6

11.2.7. $y = c \sum_{n=-\infty}^{\infty} \{H[x - n(a + b)] - H[x - n(a + b) - a]\}$
"Comb function"

1. $a = 0.03$, $b = 0.07$, $c = 0.50$

11.3. FUNCTIONS WITH A FINITE NUMBER OF DISCONTINUITIES IN FIRST DERIVATIVE

11.3.1. $y = [c/(b - a)][(x - a)H(x - a) - (x - b)H(x - b)]$
"Ramp function"

1. $a = -0.5$, $b = 0.5$, $c = 0.5$

11.3.2. $y = c(1 - |x|/a)[H(x + a) - H(x - a)]$
"Triangular function"

1. $a = 0.5$, $c = 0.5$

11.3.3. $y = c(1 - x^2/a^2)^{1/2}[H(x + a) - H(x - a)]$
"Semiellipse" ("semicircle" for $a = c$)

1. $a = 0.75$, $c = 0.50$

11.3.4. $y = c(1 - e^{-ax})H(x)$
"Exponential ramp"

1. $a = 5.0$, $c = 0.5$

11.4. FUNCTIONS WITH AN INFINITE NUMBER OF DISCONTINUITIES IN FIRST DERIVATIVE

11.4.1. $y = 2c \sum_{n=-\infty}^{\infty} \{H[(2n + 1)a/2] - H[(2n + 3)a/2]\} +$

$\sum_{n=-\infty}^{\infty} [(2cx)/a - 4nc]\{H[(2n - 1)a/2] - 2H[(2n + 1)a/2]$

$H[(2n + 3)a/2]\}$

"Triangular sine wave"

1. $a = 0.2$, $c = 0.5$

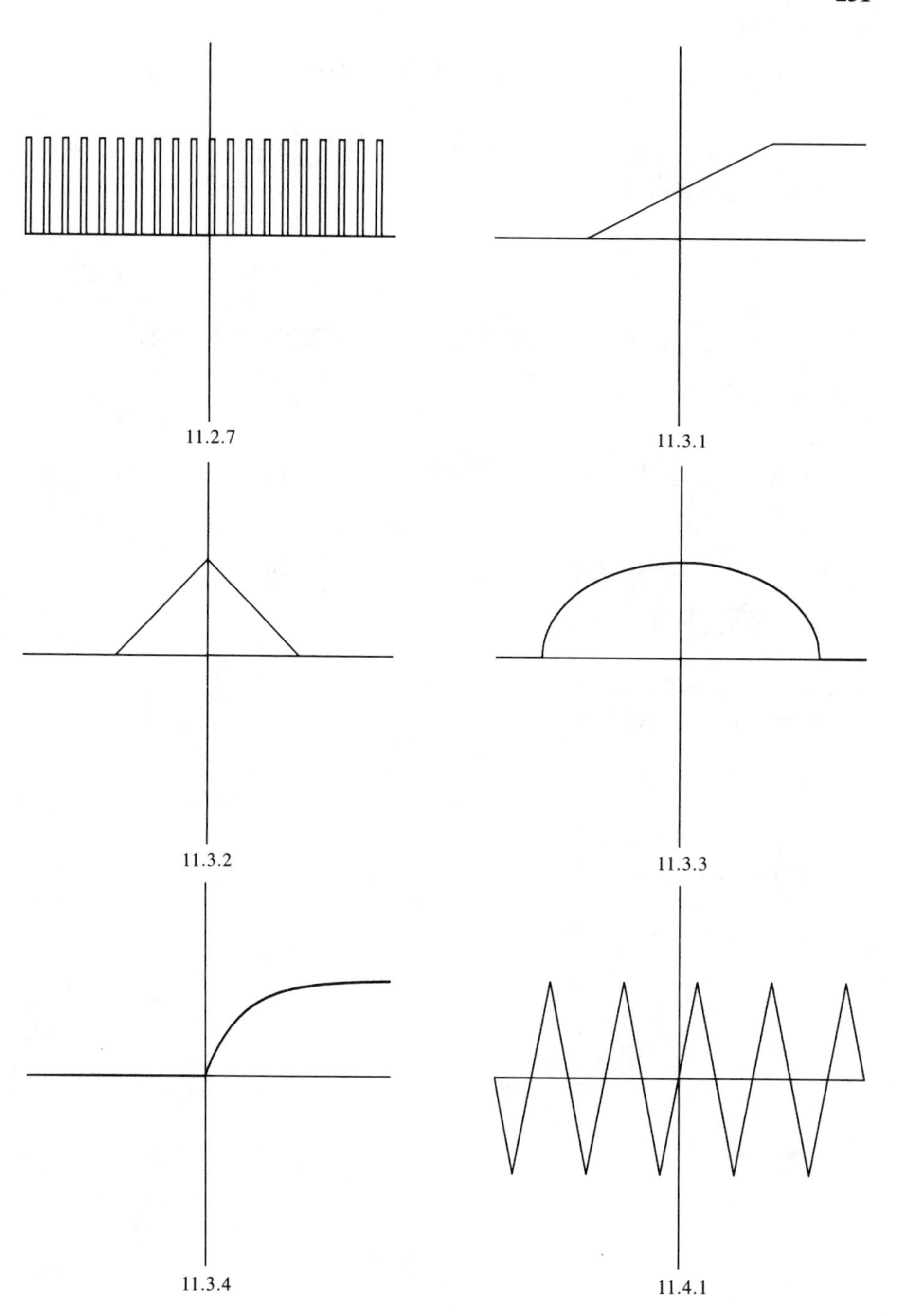
11.2.7
11.3.1
11.3.2
11.3.3
11.3.4
11.4.1

11.4.2. $y = c \sum_{n=-\infty}^{\infty} \{H[(2n - 1)a] - H[(2n + 1)a]\} +$

$$\sum_{n=-\infty}^{\infty} [(2cx)/a - 4nc]\{H[(2n - 1)a] - 2H[2na] + H[(2n + 1)a]\}$$

"Triangular cosine wave"

1. $a = 0.2$, $c = 0.5$

11.4.3. $y = \sum_{n=-\infty}^{\infty} [2cn - (cx)/a]\{H[(2n - 1)a] - 2H[2na] + H[(2n + 1)a]\}$

1. $a = 0.2$, $c = 0.5$

11.4.4. $y = \sum_{n=-\infty}^{\infty} [(cx)/a - 2cn]\{H[(2n - 1)a] - 2H[2na] + H[(2n + 1)a]\}$

$$+ c \sum_{n=-\infty}^{\infty} \{H[(2n - 1)a] - H[(2n + 1)a]\}$$

1. $a = 0.2$, $c = 0.5$

11.4.5. $y = c|\sin(2\pi ax)|$

"Rectified sine wave"

1. $a = 2.0$, $c = 0.5$

11.4.6. $y = c|\cos(2\pi ax)|$

"Rectified cosine wave"

1. $a = 2.0$, $c = 0.5$

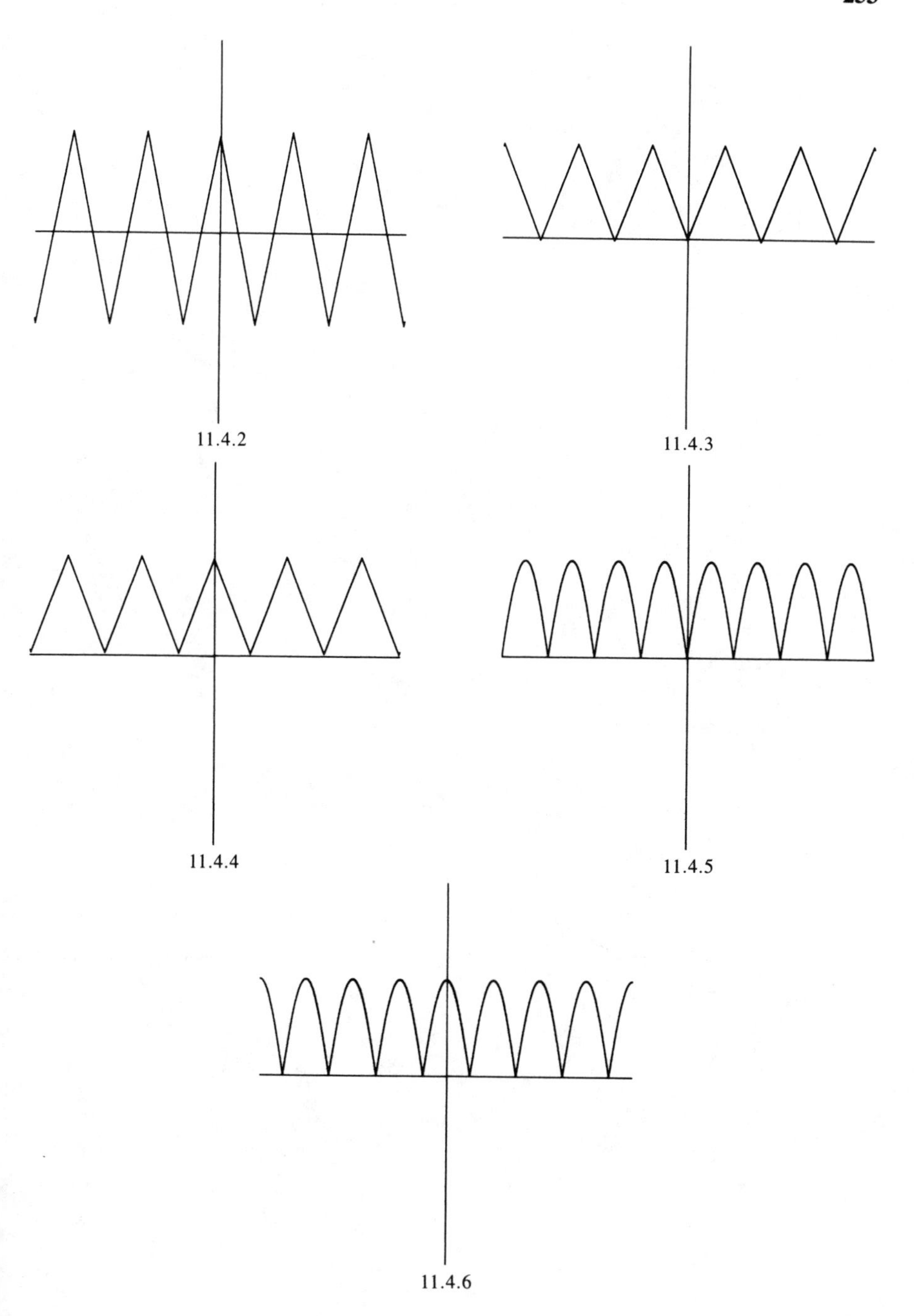
11.4.2
11.4.3
11.4.4
11.4.5
11.4.6

Chapter 12

POLYGONS

The familiar shapes of two-dimensional geometry are shown in the first three sections of this chapter. Scale is entirely relative for these figures. The last three sections show how triangles, squares, and hexagons can be combined into more complicated shapes; these can serve as building blocks for even larger patterns, and some are capable of tiling the plane.

12.1. REGULAR POLYGONS

12.1.1. Triangle (n = 3)

12.1.2. Square (n = 4)

12.1.3. Pentagon (n = 5)

12.1.4. Hexagon (n = 6)

12.1.5. Heptagon (n = 7)

12.1.6. Octagon (n = 8)

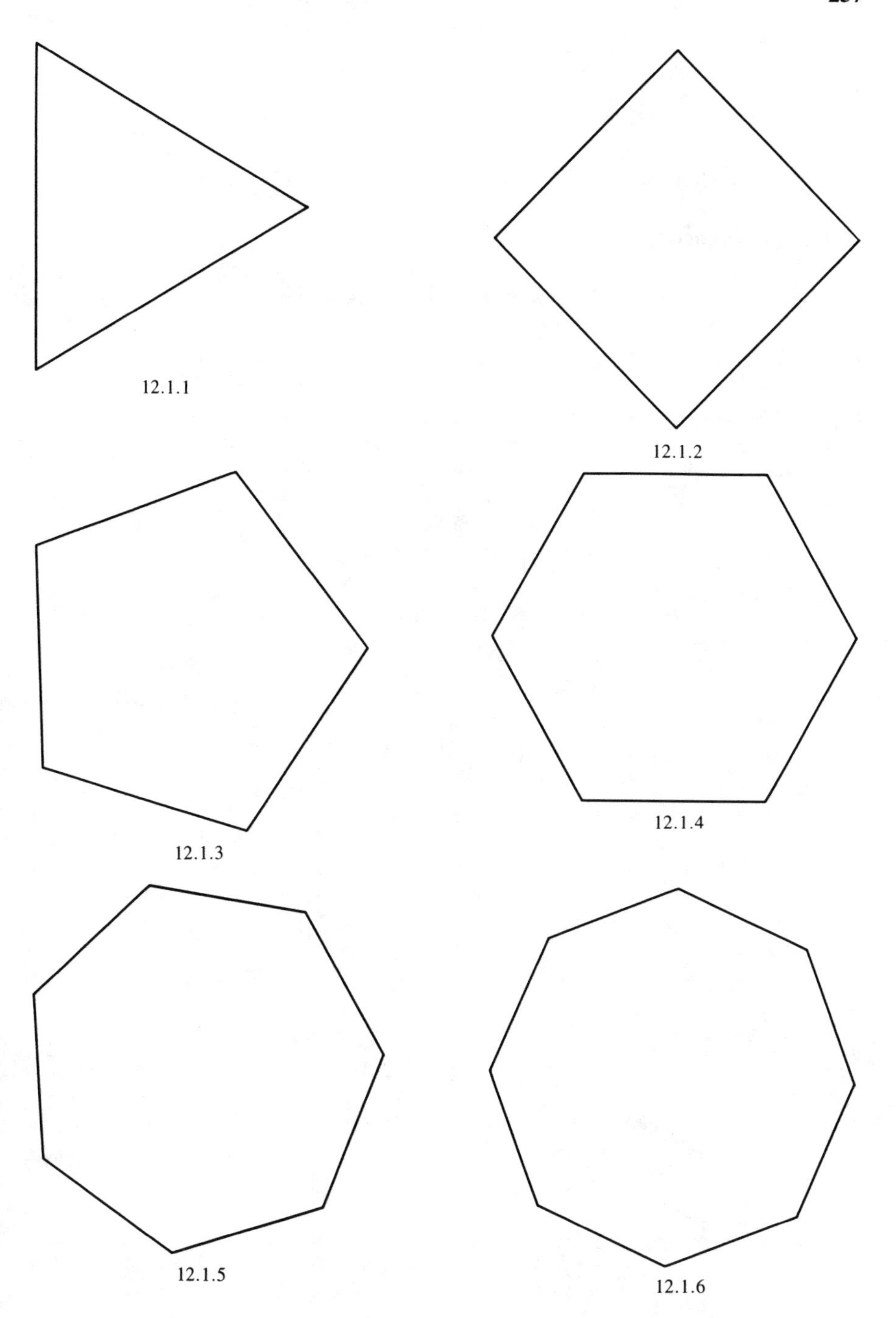

12.1.1

12.1.2

12.1.3

12.1.4

12.1.5

12.1.6

12.1.7. Nonagon (n = 9)

12.1.8. Decagon (n = 10)

12.1.9. Undecagon (n = 11)

12.1.10. Dodecagon (n = 12)

12.2. IRREGULAR TRIANGLES

12.2.1. Right triangle (one angle = 90°)

12.2.2. Isosceles triangle (two angles equal)

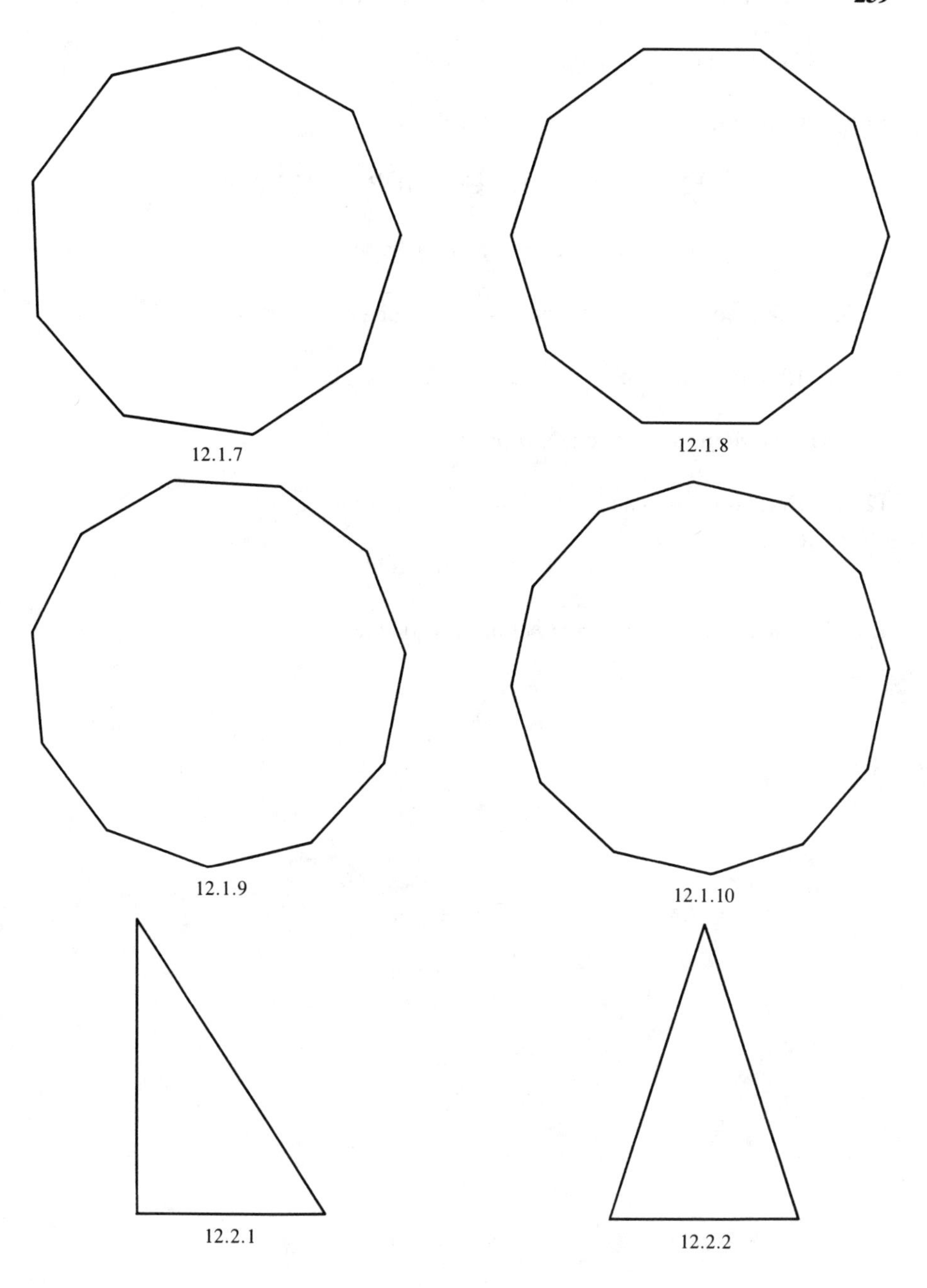

12.1.7

12.1.8

12.1.9

12.1.10

12.2.1

12.2.2

12.2.3. Acute triangle (all angles $< 90°$)

12.2.4. Obtuse triangle (one angle $> 90°$)

12.3. IRREGULAR QUADRILATERALS

12.3.1. Rectangle ($a = b$ and $c = d$ and all angles $= 90°$)

12.3.2. Parallelogram ($a = b$ and $c = d$ and angles $\neq 90°$)

12.3.3. Rhombus ($a = b = c = d$ and angles $\neq 90°$)

12.3.4. Trapezoid ($a = b$, c and d parallel)

12.3.5. Deltoid ($a = b$ and $c = d$ and two angles are equal)

12.4. POLYIAMONDS

12.4.1. Triamonds (3 connected equilateral triangles)

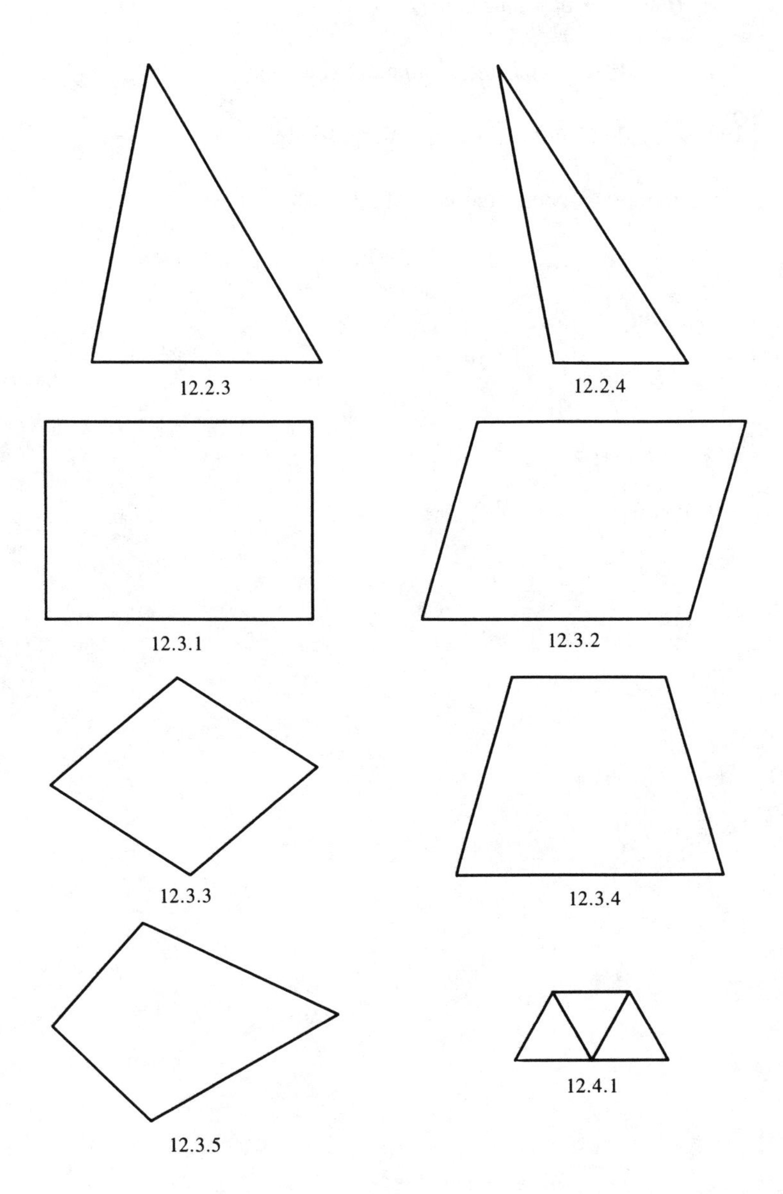
12.2.3
12.2.4
12.3.1
12.3.2
12.3.3
12.3.4
12.3.5
12.4.1

12.4.2. Tetriamonds (4 connected equilateral triangles)

12.4.3. Pentiamonds (5 connected equilateral triangles)

12.4.4. Hexiamonds (6 connected equilateral triangles)

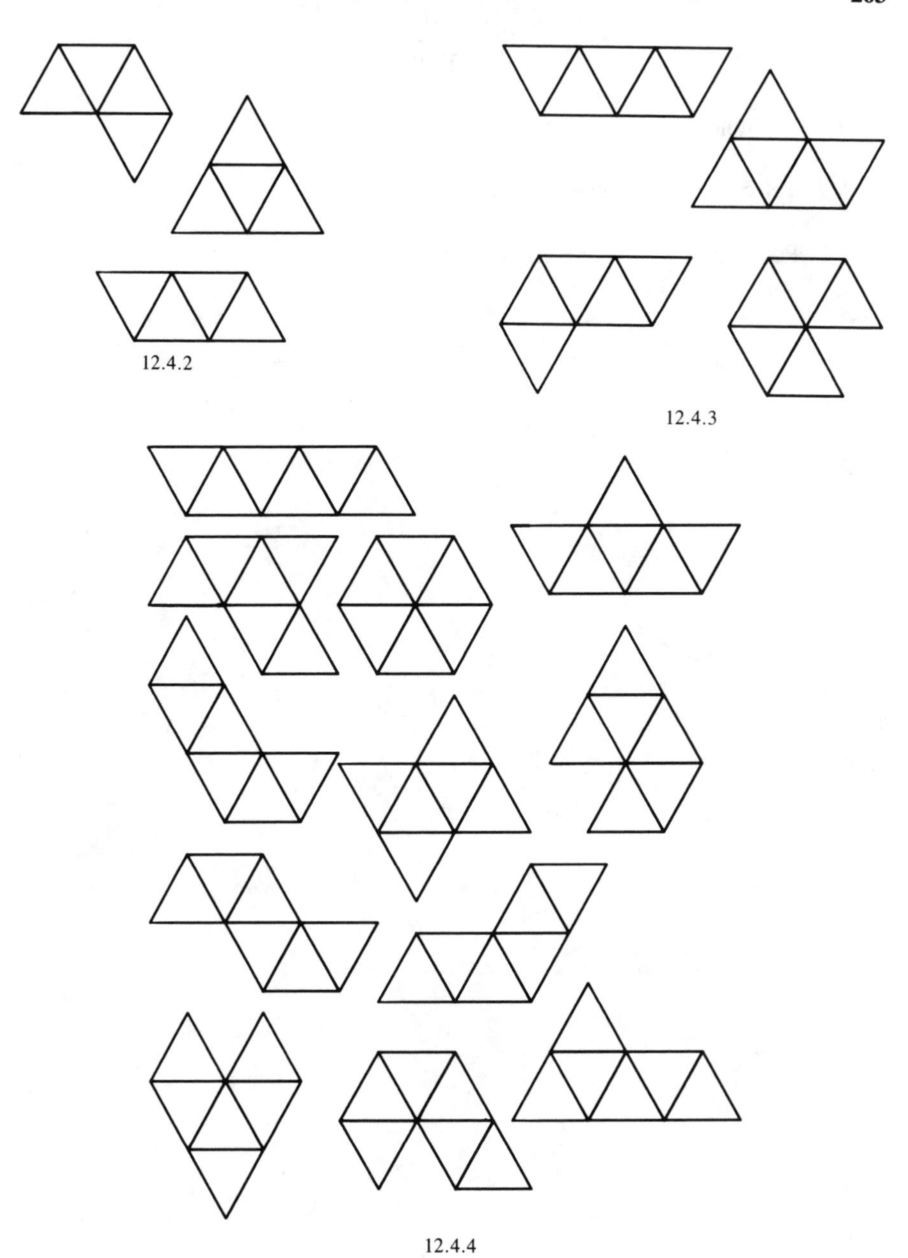

12.4.2

12.4.3

12.4.4

12.5. POLYOMINOES

12.5.1. Trominoes (3 connected squares)

12.5.2. Tetrominoes (4 connected squares)

12.5.3. Pentominoes (5 connected squares)

12.5.1

12.5.2

12.5.3

12.6. POLYHEXES

12.6.1. Trihexes (3 connected regular hexagons)

12.6.2. Tetrahexes (4 connected regular hexagons)

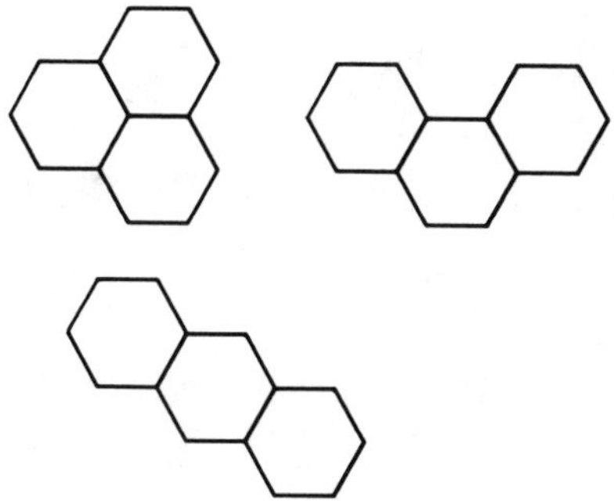

12.6.1

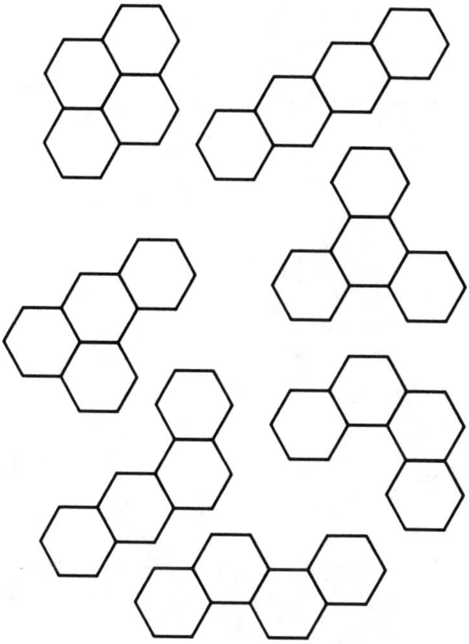

12.6.2

Chapter 13

POLYHEDRA AND OTHER CLOSED SURFACES WITH EDGES

Smooth, closed surfaces form the boundary of a volume or solid such that the outward normal of the surface at any point is everywhere continuous in all directions about that point (for example, a sphere or torus). However, some closed surfaces contain "edges", which are defined to be where the derivative of the surface (and therefore the normal) is discontinuous. The closed surfaces of this chapter will be represented by the lines or curves which constitute their edges. Any hidden edges which would not be visible to the observer, who is viewing the surface from infinity, are represented by dotted lines while visible edges are solid lines. The projection used here is again the orthographic parallel one discussed in the text in Chapter 8. The rotation angle θ is not given for the figures here because it is not meaningful in this context. The viewing inclination ϕ from the vertical axis is given though; its meaning derives from the fact that for each figure either the base or a plane of symmetry is coincident with the (x,y) plane.

13.1. REGULAR POLYHEDRA

13.1.1. Tetrahedron ($n = 4$) ($\phi = 70$)

13.1.2. Hexahedron ($n = 6$) ($\phi = 70$)

13.1.3. Octahedron ($n = 8$) ($\phi = 70$)

13.1.4. Dodecahedron ($n = 12$) ($\phi = 70$)

13.1.5. Icosahedron ($n = 20$) ($\phi = 70$)

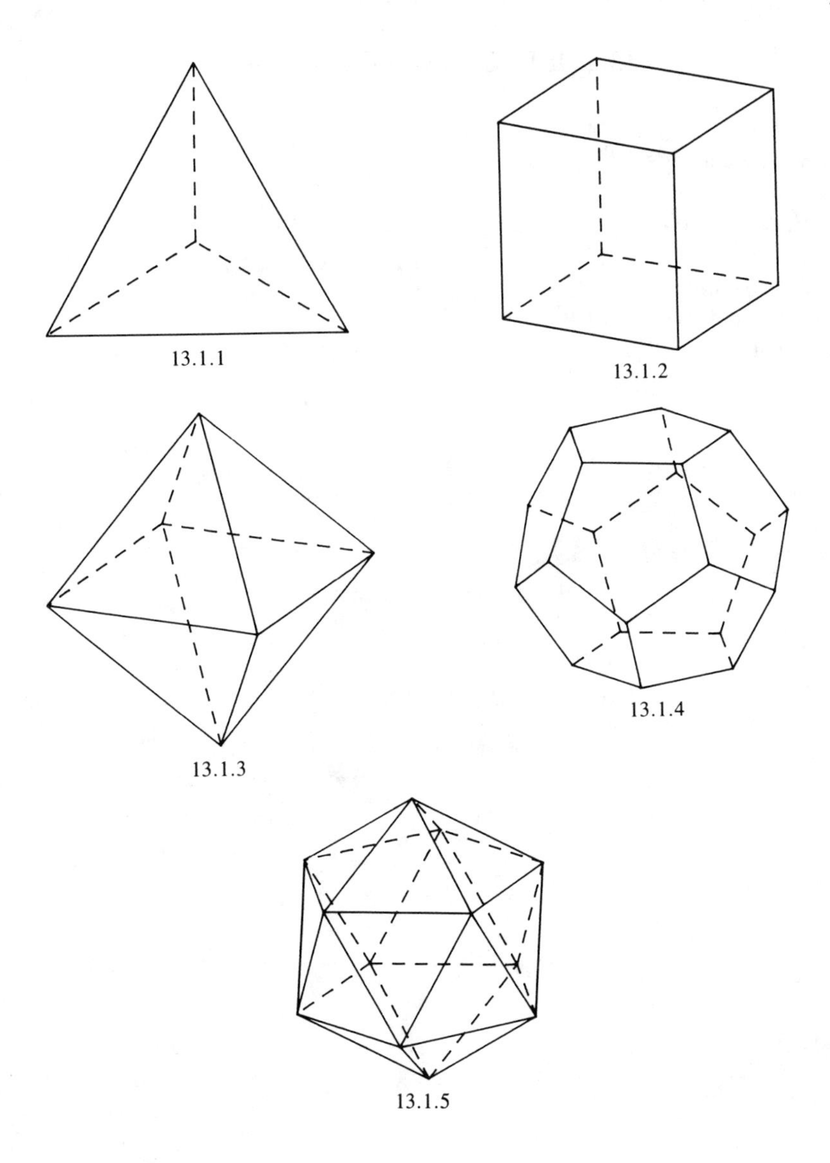

13.1.1

13.1.2

13.1.3

13.1.4

13.1.5

13.2. IRREGULAR POLYHEDRA

13.2.1. Prism

1. Triangular ($\phi = 70$)
2. Square ($\phi = 70$)
3. Hexagonal ($\phi = 70$)

13.2.2. Prismoid

1. Triangular ($\phi = 70$)
2. Square ($\phi = 70$)

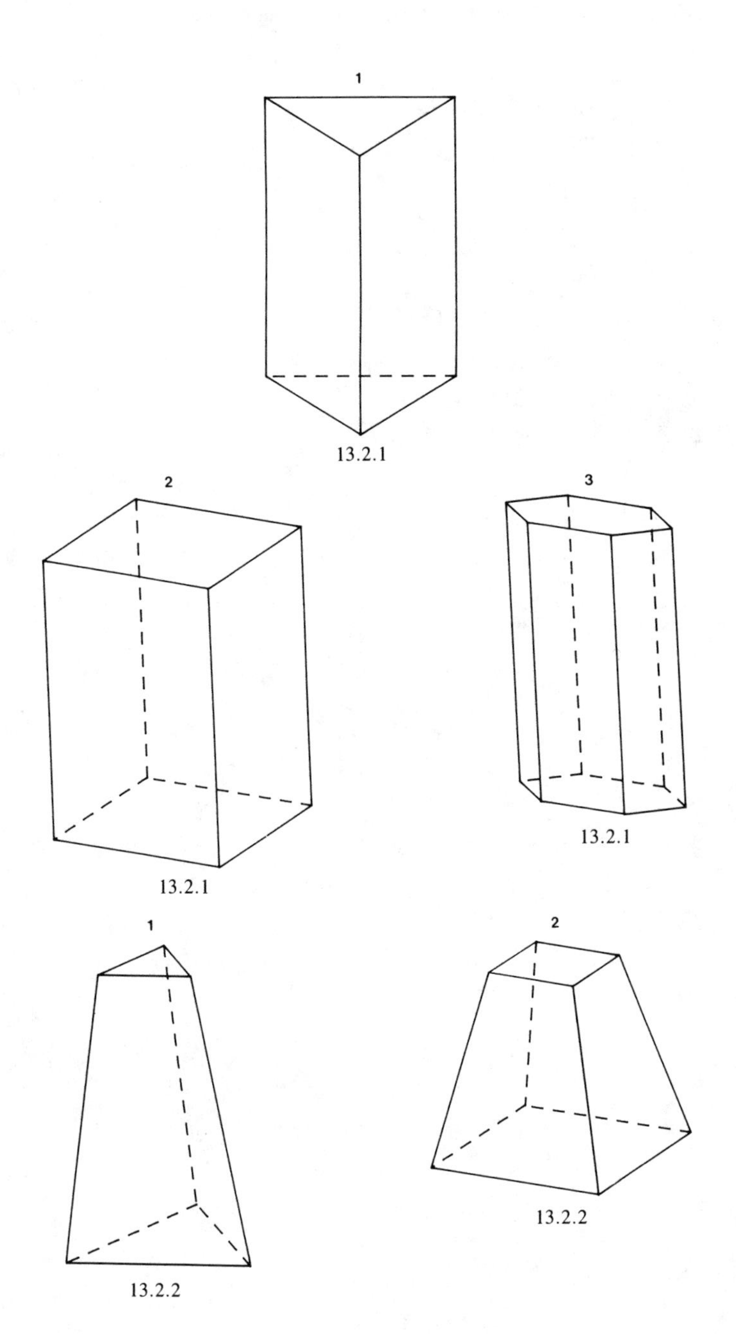

13.2.1

13.2.1

13.2.1

13.2.2

13.2.2

13.2.3. Prismatoid

1. Triangular-hexagonal ($\phi = 70$)
2. Square-octagonal ($\phi = 70$)

13.2.4. Parallelepiped

1. Oblique ($\phi = 70$)
2. Right ($\phi = 70$)

13.2.5. Pyramid

1. Regular — triangle base ($\phi = 70$)
2. Regular — square base ($\phi = 70$)
3. Irregular — rhombus base ($\phi = 70$)

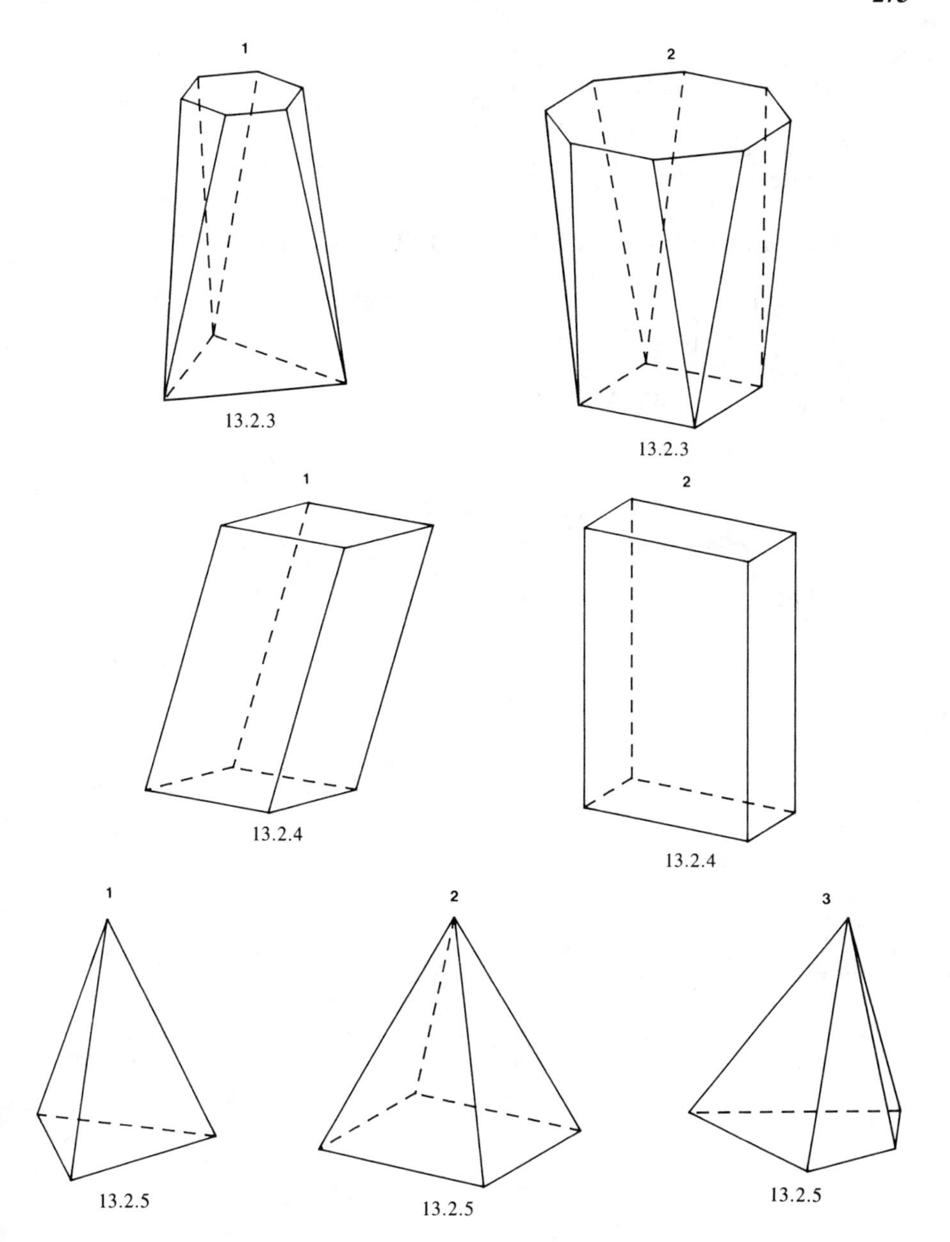
1
2
13.2.3
13.2.3
1
2
13.2.4
13.2.4
1
2
3
13.2.5
13.2.5
13.2.5

13.2.6. Dipyramid
1. Regular — square base ($\phi = 70$)

13.2.7. Trapezohedron
1. Regular — 8 sides ($\phi = 70$)

13.3. MISCELLANEOUS CLOSED SURFACES WITH EDGES

13.3.1. Cylinder
1. Right circular ($\phi = 70$)
2. Oblique circular ($\phi = 70$)
3. Right circular (disk) ($\phi = 70$)

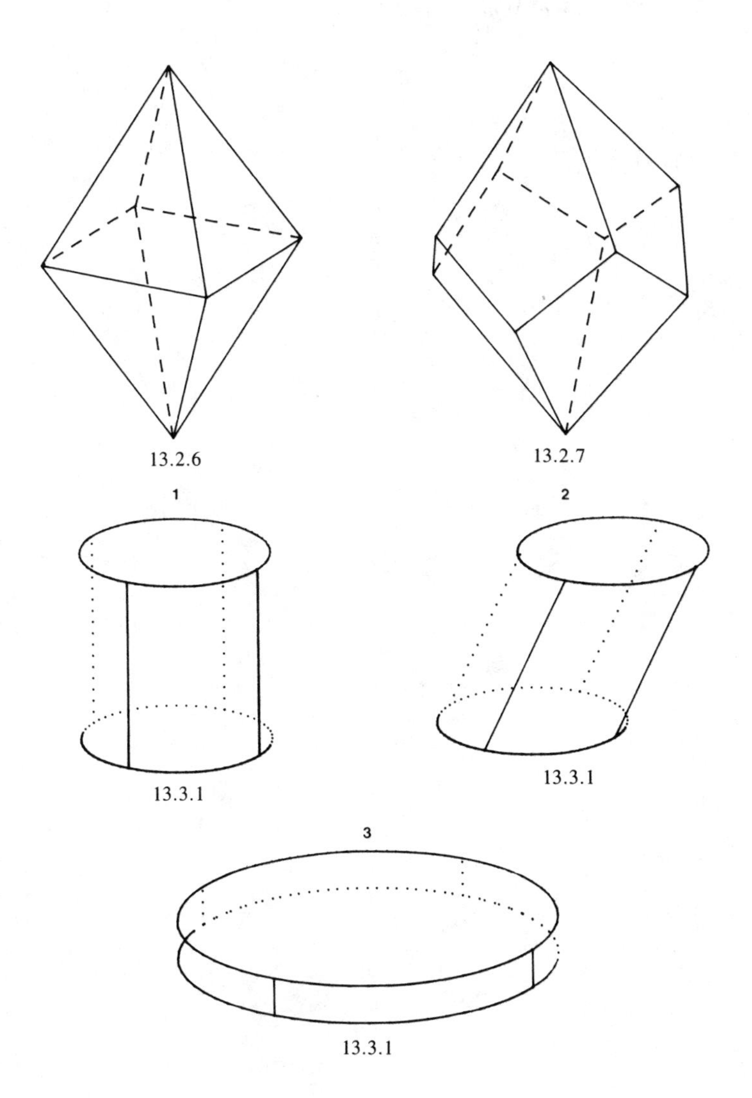

13.2.6

13.2.7

13.3.1

13.3.1

13.3.1

13.3.2. Cone
1. Right circular ($\phi = 70$)
2. Oblique circular ($\phi = 70$)
3. Frustrum ($\phi = 70$)

13.3.3. Hemisphere ($\phi = 70$)

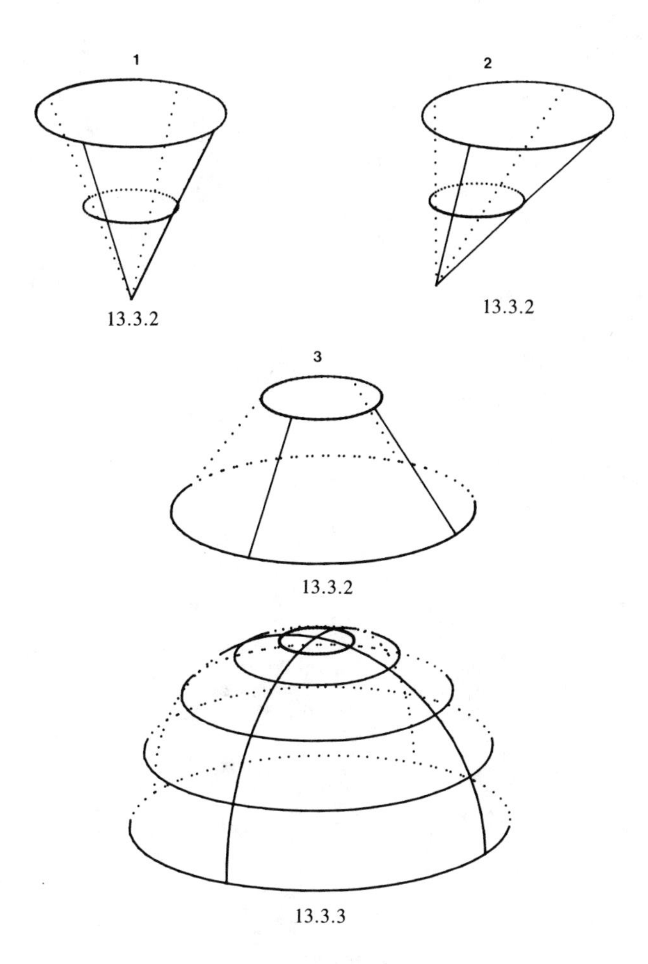

13.3.2

13.3.2

13.3.2

13.3.3

INDEX

D

E

F

G

S

T

U

W

Z